ALLE · ZEIT · WACH · 1842

Guido di Prisco (Ed.)

Life Under Extreme Conditions

Biochemical Adaptation

With 43 Figures

Springer-Verlag
Berlin Heidelberg New York
London Paris Tokyo
Hong Kong Barcelona

Professor Dr. Guido di Prisco
Institute of Protein Biochemistry
and Enzymology (IBPE)
Italian National Research Council
Via Marconi 10
I-80125 Naples

Cover illustration see Fig. 5, page 11

ISBN 3-540-53108-4 Springer-Verlag Berlin Heidelberg New York
ISBN 0-387-53108-4 Springer-Verlag New York Berlin Heidelberg

Library of Congress Cataloging-in-Publication Data. Life under extreme conditions: biochemical adaptation/Guido di Prisco, ed. p. cm. Papers from a session of the 19th FEBS Meeting, held in Rome, July 2-7, 1989. Includes bibliographical references and index. ISBN 3-540-53108-4 (Springer-Verlag Berlin Heidelberg New York: acid-free paper).– ISBN 0-387-53108-4 (Springer-Verlag New York Berlin Heidelberg: acid-free paper) 1. Adaptation (Physiology) – Congresses. 2. Extreme environments – Congresses. I. Di Prisco, Guido, 1937. II. Federation of European Biochemical Societies. Meeting (19th: 1989: Rome, Italy) QP82.L54 1991 574.5 – dc20 90-10372 CIP

Printed in Germany

Typesetting: International Typesetters Inc., Makati, Philippines
31/3130-543210 – Printed on acid-free paper

Preface

The scientific program of the 19th FEBS (Federation of European Biochemical Societies) Meeting (Rome, July 2–7, 1989) included a session on "Biochemistry under Extreme Life Conditions". The lectures brought together a large number of scientists engaged in shedding light on the many mechanisms through which living organisms achieve molecular adaptation to the rather exceptional conditions occurring in several domains of this planet. These conditions include low and high temperatures and high concentration of salts. The possibility of exchanging ideas and of establishing interactions in fields seemingly remote from one another in the context of an international biochemistry congress of primary importance was greeted with satisfaction, also with consideration that such a "proximity of extremes" had actually been witnessed on only a few occasions in the past. In fact, the development and steady growth of research on the adaptation at the molecular level of polar organisms to environmental conditions are relatively recent events, and have originated in the increasing number of countries whose governments have decided to engage in polar science programs.

Antarctica, possibly the only part of the earth whose natural habitat is as yet largely immune from mankind's disturbance, is consequently a giant, natural laboratory. It differs from northern polar regions in many respects. Following the fragmentation of the supercontinent Gondwana (of which it was a part) and the continental drift, the subsequent cooling of the Antarctic environment gradually changed the once temperate land into a bitterly cold, dry, ice-covered desert which permits life merely along the coastline. In general, the temperature of the Antarctics constantly close to −1.9 °C, the equilibrium temperature of the ice-seawater mixture. The Antarctic Convergence (a well-defined, circular, oceanic frontal system, running between 50° S and 60 °S) acts as a natural barrier to marine organisms in both directions. South of the Convergence, fish have remained in isolation during the past 25 million years or more. The need to adapt to the progressive cooling conditioned the evolution, and Antarctic fish now tolerate conditions that would be prohibitive to fish from temperate waters. In fact, these fish can survive only under these conditions, since exposure to water temperatures of a few degrees above zero has lethal effects.

Some of the specializations achieved through the evolutionary trends during cold adaptation are illustrated in three of the contributions, dealing with (1) freezing resistance (in contrast to Arctic fish, which produce "antifreeze" proteins only seasonally, Antarctic teleosts need these molecules in the interstitial fluids all year-round); (2) cold-stable microtubules (tubulins assemble and form microtubules at temperatures that in warm-blooded organisms favor dissociation); and (3) the biochemistry of oxygen transport (the hematological adaptation of Antarctic fish compensates for the increase in blood viscosity caused by low temperatures; the colorless blood of the species of one family has neither hemoglobin nor erythrocytes).

An ideal link to the mechanism of respiration in polar fish may be found in the functional properties of hemoglobins of Arctic mammals that live at very low temperatures (and that may experience enormous seasonal temperature variations) but are not ectotherms. In general, the release of oxygen is thermodynamically impaired when the temperature of tissues decreases. However, in reindeer, musk-oxen and whales, the unusual temperature dependence of the oxygen-binding curve of hemoglobin indicates an exception to this rule. Oxygen unloading in peripheral tissues, where temperature may be lower than that of the lungs, will not be impaired; this feature may be regarded as being among the molecular adaptations to extreme environmental conditions.

Moving away from the chilly habitats, the present survey takes the reader into environments resembling the much warmer ones where primitive life found its origin.

Physiological, morphological, and biochemical analysis has led in the 1970s to the classification of prokaryotes into two phylogenetically coherent groups: eubacteria and archaebacteria. The ecological niches of the latter group are characterized by one or more "extreme" factors such as anaerobiosis, high ionic strength, extreme pH values, and temperatures reaching 140 °C. The main phenotypes of archaebacteria comprise (hyper) thermophilic and halophilic bacteria. The former are perhaps the most ancient, the ancestral form of life. They appeared when the earth was still very warm and there was no oxygen, and grow anaerobically at temperatures often higher than 110 °C. Halophilic bacteria, which appeared over 1500 million years ago, grow aerobically and require salt concentrations as high as 4 M.

The other chapters of this volume deal with various aspects of the molecular adaptations of these microorganisms. A review of lipid structure and biosynthesis yields valuable information on the role of these molecules in providing the ability to withstand environmental stresses to the membranes of archaebacteria. Investigations on the relationship between the molecular structure and the unusual stability displayed by enzymes of extremophiles also give clues to identifying the optimal exploitation of the latter property by modern biotechnology in industrial processes. Finally, the implications of the stability features of halobacterial malate dehydrogenase in formulating a general model for the stabilization of halophilic proteins under conditions of extreme salinity are discussed.

Research focused on so many forms of molecular adaptation to life under extreme conditions, which differ so drastically from one another, once again emphasizes the endless ability of nature to modulate evolution in such a way that living organisms will finally find their way to growth and reproduction, no matter how challenging the environmental conditions may be. As indicated by the articles in this volume, macromolecules (enzymes and other proteins and glycoproteins, lipids, nucleic acids), through the relationships between their molecular structure and biological function and through gene expression, are tools as well as products of the various adjustments. Research in this field, in conjunction with genetics and physiology, will certainly provide precious insights into the history of evolution.

I am thoroughly confident that this volume will faithfully reflect the interesting and significant contributions along these lines, successfully offered by the speakers during the session of the 19th FEBS Meeting.

Naples, Autumn 1990 GUIDO DI PRISCO

Contents

List of Contributors

You will find the addresses at the beginning of the respective contribution

The Role of Antifreeze Glycopeptides and Peptides in the Freezing Avoidance of Cold-Water Fish

C.C. CHENG[1] and A.L. DEVRIES[1]

1 Introduction

The Antarctic and Arctic oceans are perennially at about –1.9 °C, the freezing point of seawater, and ice-covered in their shallow waters (Littlepage 1965). The near-shore waters of the north temperate oceans reach this freezing temperature during the winter seasons. This is well below the freezing point of between –0.5 to –0.9 °C of a typical marine teleost (Black 1951). In the presence of ice, supercooling is impossible, and therefore a 1 °C difference between the freezing point of the fish's body fluids and the environment would lead to freezing. Freezing or even partial freezing has been shown in all cases to result in death (Scholander et al. 1957). However, many fish living in these freezing environments frequently come into contact with ice (DeVries and Wohlschlag 1969; DeVries 1970, 1971, 1974), yet do not appear to freeze (Hargens 1972). In fact, some use the abundant ice crystal formations as a habitat and spend their entire lives there to forage for food and to escape predators (Andriashev 1970; DeVries and Lin 1977a). Freezing in these fish does not occur unless they are exposed to temperatures below –2.2 °C in the presence of ice (Scholander et al. 1957; DeVries and Lin 1977a).

In most temperate marine fish, sodium chloride is the principle blood serum solute and is responsible for 85% of the observed freezing point depression (Gordon et al. 1962). The remainder of the freezing point depression is due to small amounts of potassium, calcium, urea, glucose, and the free amino acids (Potts and Parry 1964). In fish inhabiting freezing environments, concentrations of sodium chloride in the body fluids are somewhat higher relative to temperate forms (Fletcher 1977; O'Grady and DeVries 1982; Ahlgren et al. 1988). However, they in fact are responsible for only 40–50% of the observed freezing point depression (Duman and DeVries 1975; Petzel et al. 1980). The additional freezing point depression was first shown in the antarctic fish to be associated with solutes in the colloidal fraction of the serum; over half of the blood freezing point depression is retained by a dialysis membrane with a molecular weight cut-off of 3000 daltons (DeVries 1974, 1980, 1982). In the Antarctic nototheniid fish and in some northern cods these solutes are relatively large glycopeptides, while in the Antarctic eel pouts and a number of northern fish, they are peptides. Antifreeze

[1] Department of Physiology and Biophysics, 524 Burrill Hall, 407 S. Goodwin Avenue, University of Illinois, Urbana, IL 61801, USA.

Guido di Prisco (Ed)
Life Under Extreme Conditions

glycopeptides appear in at least eight sizes with molecular weights ranging between 2600 and 34 000 daltons (DeVries et al. 1970; DeVries 1974; Feeney and Yeh 1978), whereas the peptides occur in fewer size classes and range between 3200 and 14000 daltons, depending on the species (Duman and DeVries 1976; Slaughter et al. 1981; DeVries 1982, 1986). On a molal basis, the antifreezes depress the freezing point by 200 – 300 times more than that expected on the basis of colligative relations alone (DeVries 1971). These glycopeptides and peptides lower only the freezing point noncolligatively, but show the expected colligative effect on the melting point of the solid phase (ice). The noncolligative lowering of the freezing point has been referred to as an antifreeze effect and these molecules are referred to as “antifreezes”.

2 Freezing Behavior of Fish Body Fluids Containing Antifreeze

Experimentally, the melting point of a solution can be accurately obtained by determining the temperature at which a small polycrystalline ice crystal (~50 μm diam.) begins to melt as the temperature of the solution is raised slowly (0.01 °C/min). If the temperature is lowered very slightly before the ice crystal is completely melted, the crystal can be observed to increase slowly in size. This temperature, or the freezing point of the solution, differs from the melting point by only 1/100 – 2/100 of a degree, i.e., they are essentially the same. Freezing points determined in this manner are referred to as equilibrium freezing points, which are the equivalent of the melting points (DeVries 1986).

Using this technique, the melting point of a seed crystal in Antarctic fish serum is found to be about –1 °C, accountable largely by the amount of salts present (O’Grady and DeVries 1982). However, no growth of the seed crystal is observed until the temperature is lowered to –2.2 °C. In contrast to the slow dendritic crystal growth that occurs in salt and other biological solutions, growth from the seed crystal in Antarctic fish serum is very rapid, and is in the form of long narrow spicules resembling strands of glass wool (DeVries 1971; Raymond and DeVries 1977). The blood freezing point corresponds, within a few tenths of a degree, to the freezing temperature of the whole animal. When the serum is dialyzed to remove salts and other small solutes, the melting point becomes –0.02 °C, and the freezing point, –1.2 °C. This unusually large separation of melting and freezing points, or thermal hysteresis, is due to marcromolecular antifreeze in the fish serum retained within the dialysis membrane (DeVries and Wohlschlag 1969; DeVries 1971; Raymond et al. 1975; Duman and DeVries 1976). The freezing point and the melting point of the dialyzed serum of a temperate fish are both –0.01 °C, indicating no antifreeze is present. The melting point-freezing point difference method has been widely used to determine the presence and concentration of antifreeze in blood serum or dialyzed serum (DeVries 1974).

3 Antifreeze Glycopeptides and Peptides

Two types of antifreeze have been isolated from polar and north-temperate fishes, antifreeze glycopeptides (AFGPs) or peptides (AFPs). AFGPs were first discovered and isolated from the Antarctic nototheniid fishes which comprise the largest family of fish in the Antarctic ocean (DeVries and Lin 1977a; Schneppenheim and Theede 1982; Ahlgren and DeVries 1984). These AFGPs have been thoroughly characterized. They are composed of repeating units of the tripeptide alanyl-alanyl-threonine with the disaccharide β-D-galactopyranosyl-(1 → 3)-2-acetamido-2-deoxy-α D-galacto-pyranose linked to threonine residues (Fig. 1; DeVries et al. 1970; Komatsu et al. 1970; DeVries et al. 1971; Shier et al. 1972, 1975). There are at least eight different sizes, depending on the number of repeats of the glycotripeptide unit, and the range of molecular weights is between 2600 and 34 000 daltons (Table 1). The three smaller AFGPs differ from the larger ones in that proline replaces some of the alanines (Lin et al. 1972; Morris et al. 1978). The same eight AFGPs have also been isolated from a northern gadid in the Labrador, the rock cod, *Gadus ogac* (Van Voorhies et al. 1978). Similar ones have been reported to be present in some of the other high-latitude northern cods belonging to the family Gadidae (Raymond et al. 1975; Osuga and Feeney 1978).

Antifreezes peptides (AFPs) have been isolated from a number of northern fish and two Antarctic eel pouts. Three structural groups are recognizable. The AFPs of winter flounder, Alaskan plaice (flat fish), and short-horn sculpin (cottid) have molecular weights between 3000 to 4000 daltons, and are primarily al-

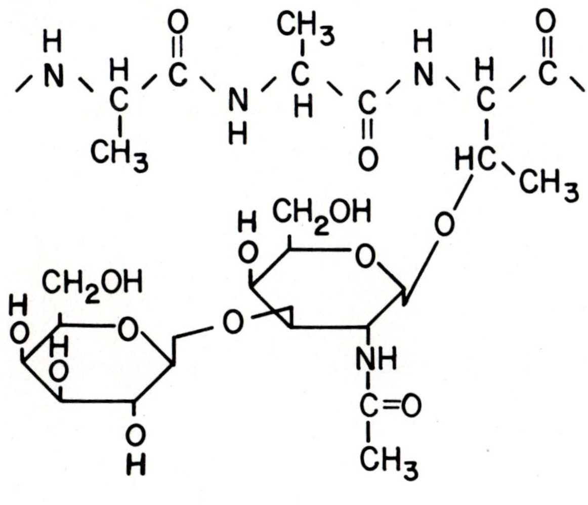

Fig. 1. Basic repeating structural unit of the antifreeze glycopeptides isolated from the blood of Antarctic nototheniid fish. The peptide is made up of only two amino acids, alanine and threonine in the sequence alanyl-alanyl-threonine. Every threonine is joined by a glycosidic linkage to the disaccharide β-D-galactopyranosyl-(1 → 3)-2-acetamido-2-deoxy-α-D-galactopyranose. The small glycopeptides have proline replacing a few of the alanines, but otherwise are the same as the large glycopeptides

Table 1. Molecular weights of antifreeze glycopeptides isolated from the Antarctic nototheniid *Pagothenia borchgrevinki*

Antifreeze glycopeptides	Molecular weight[a]
1	33 700
2	28 800
3	21 500
4	17 000
5	10 500
6[b]	7900
7	3500
8	2600

[a] Molecular weights were determined by sedimentation equilibrium centrifugation.
[b] Antifreeze glycopeptide 6 separates into four different bands on acrylamide gel electrophoresis.

pha-helical peptides of repeating stretches of nonpolar alanines separated by short segments of polar residues, as shown in Fig. 2 (DeVries and Lin 1977b; Gourlie et al. 1984; Hew et al. 1985; DeVries, unpublished). AFPs of sea raven, a cottid, are about 14 000 daltons, contain substantial numbers of cysteines, and possess beta-structure (Slaughter et al. 1981; Ng et al. 1986). The AFPs from the zoarcid fish, which include the Atlantic ocean pout (Li et al. 1985), an Arctic and two Antarctic eel pouts (Schrag et al. 1987; Cheng and DeVries 1989), comprise

Winter Flounder:

DTASDAAAAAALTAANAAAAAKLTADNAAAAAAATAA

Alaskan Plaice:

DTASDAAAAAAATAAAAKAAAEKTARDAAAAAAATAAAAR

Short−horn Sculpin:

MDGETPAQKAARLAAAAAALAAKTAADAAAKAAAIAAAAASA

Fig. 2. Amino acid sequences of the helical antifreeze peptides. The 11-residue repeats are indicated by *brackets* (see text)

a third class of AFPs. The zoarcid AFPs are about 7000 daltons in size, and unlike any known antifreezes, they have no biased amino acid composition, no repeating sequences in their primary structure, and their secondary structures appear to be random chains (Ananthanarayanan et al. 1986; Schrag et al. 1987; Cheng and DeVries 1989). Interestingly, despite the vast geographic separation of the zoarcids, their AFPs are highly conserved in amino acid sequence (Fig. 3), with sequence homologies between 60 and 85% (Cheng and DeVries 1989). Peptide heterogeneity exists in the AFPs in that each fish synthesizes more than one AFP of similar size, with minor variation in the amino acid composition.

4 Fluid and Tissue Distribution of Antifreeze in the Fish

The AFGPs and AFPs constitute as much as 3% of blood protein in many of these cold-water fish, and along with sodium chloride, they depress the fish's freezing points below that of seawater (–1.9 °C). The antifreezes are synthesized by the liver (Hudson et al. 1979; O'Grady et al. 1982a) and are secreted into the circulatory system, where they then become distributed to the other fluid compartments of the fish; no antifreeze is found intracellularly (Ahlgren et al. 1988).

```
           1                                           10                                          20
AB         THR-LYS-SER-VAL-VAL-ALA-SER-GLN-LEU-ILE-PRO-ILE-ASN-THR-ALA-LEU-THR-PRO-ALA-MET-
       1                                           10                                          20
RD     ASN-LYS-ALA-SER-VAL-VAL-ALA-ASN-GLN-LEU-ILE-PRO-ILE-ASN-THR-ALA-LEU-THR-LEU-ILE-MET-
LP     ASN-LYS-ALA-SER-VAL-VAL-ALA-ASN-GLN-LEU-ILE-PRO-ILE-ASN-THR-ALA-LEU-THR-LEU-VAL-MET-
                1                                           10
MA              GLN-SER-VAL-VAL-ALA-THR-GLN-LEU-ILE-PRO-ILE-ASN-THR-ALA-LEU-THR-PRO-ALA-MET-

           21                                      30                                          40
AB         MET-LYS-ALA-LYS-GLU-VAL-SER-PRO-LYS-GLY-ILE-PRO-ALA-GLU-GLU-MET-SER-LYS-ILE-VAL-
                                                   30                                          40
RD         MET-LYS-ALA-GLU-VAL-VAL-THR-PRO-MET-GLY-ILE-PRO-ALA-GLU-ASP-ILE-PRO-ARG-ILE-ILE-
LP         MET-ARG-ALA-GLU-VAL-VAL-THR-PRO-ALA-GLY-ILE-PRO-ALA-GLU-ASP-ILE-PRO-ARG-LEU-VAL-
           20                                      30
MA         MET-GLU-GLY-LYS-VAL-THR-ASN-PRO-ILE-GLY-ILE-PRO-PHE-ALA-GLU-MET-SER-GLN-ILE-VAL-

           41                                      50                                          60
AB         GLY-MET-GLN-VAL-ASN-ARG-ALA-VAL-ASN-LEU-ASP-GLU-THR-LEU-MET-PRO-ASP-MET-VAL-LYS-
                                               50                                              60
RD         GLY-MET-GLN-VAL-ASN-ARG-ALA-VAL-PRO-LEU-GLY-THR-THR-LEU-MET-PRO-ASP-MET-VAL-LYS-
LP         GLY-LEU-GLN-VAL-ASN-ARG-ALA-VAL-LEU-ILE-GLY-THR-THR-LEU-MET-PRO-ASP-MET-VAL-LYS-
           40                                      50
MA         GLY-LYS-GLN-VAL-ASN-THR-PRO-VAL-ALA-LYS-GLY-GLN-THR-LEU-MET-PRO-ASN-MET-VAL-LYS-

           61      63
AB         THR-TYR-GLN
                   64
RD         ASN-TYR-GLU      66
LP         GLY-TYR-ALA-PRO-GLN
           60                  65
MA         THR-TYR-VAL-ALA-GLY-LYS
```

Fig. 3. Amino acid sequences of the antifreeze peptides from the zoarcid fish. *AB* Antarctic eel pouts (*Austrolycichthys brachycephalus*), *RD* (*Rhigophila dearborni*); *LP* Arctic eel pout (*Lycodes polaris*); *MA* Atlantic ocean pout (*Macrozoarces americanus*). Sequences are positioned to demonstrate the substantial amount of homology

4.1 Freezing Avoidance of Fluids Containing Antifreeze

In the Antarctic nototheniid fish, the pericardial, peritoneal, and extradural fluid all contain AFGPs 1–8, at comparable or somewhat lower concentrations than those found in blood (Ahlgren et al. 1988). The distribution of AFGPs into these fluid compartments is passive, possible by diffusion through the capillary pores down their concentration gradients as in many of the other secreted blood proteins (Ahlgren et al. 1988).

Marine fish maintain their salt and water balance by drinking seawater, absorbing NaCl and water in the intestine, and "pumping out" the excess NaCl through the gills (Karnaky 1986). Since much of the polar waters are ice-laden, the intestinal fluids in the polar fish are in danger of freezing. In the Antarctic nototheniid fish, the intestinal fluids are fortified with AFGPs 7 and 8 and some AFGP 6 (O'Grady et al. 1983). Recent studies indicate that the AFGPs are translocated from the blood to the bile via a paracellular route (DeVries, unpublished). They enter along with the bile into the anterior end of the intestine when the gallbladder evacuates. The AFGPs are neither digested nor reabsorbed during their transit through the intestinal tract. As salt, water, and digested foodstuffs are absorbed, the intestinal fluid decreases in volume and therefore becomes concentrated with respect to AFGPs (O'Grady et al. 1982b). Fluid collected at the distal part of the intestinal tract has a freezing point of −2.2 °C (O'Grady et al. 1983), a temperature below that of the environment.

4.2 Freezing Avoidance of Fluids Lacking Antifreeze

The endolymph and the urine have no detectable amounts of antifreeze (Dobbs et al. 1974; Ahlgren et al. 1988). Endolymph is produced in the semicircular canals by secretion and therefore contains no proteins. Ice propagation into endolymph is unlikely because of its deep-seated location. In addition, the surrounding tissues are fortified with antifreeze.

Urine is antifreeze-free through different mechanisms in different fish. Antarctic nototheniid fish kidneys are composed of aglomerular nephrons, therefore AFGPs do not appear in the urine because no filtration is involved in urine formation. A survey of the nototheniid fish indicates that all species with AFGPs have aglomerular nephrons, while those lacking antifreeze because they live in warmer water have glomerular nephrons (Dobbs et al. 1974; Dobbs and DeVries 1975b; Eastman and DeVries 1986). The Antarctic eel pouts have glomerular kidneys, and their AFPs are small enough to filter through. However, no filtration occurs because the filtration barrier is very thick, and the Bowman's space does not appear to be connected with the tubule lumen (Eastman et al. 1979). In winter flounder, urine formation does involve filtration, but no AFPs appear in the urine because a charge-charge repulsion mechanism prevents these acidic peptides from crossing the anionic filtration barrier, thus keeping them in circulation (Boyd and DeVries 1983). It is important that these cold-water fish conserve their antifreezes, since continuous, rapid loss via urine would be energetically very costly.

The ocular fluids in the nototheniid fish contain small amounts of the small AFGPs (Turner et al. 1985; Ahlgren et al. 1988), but insufficient to depress the freezing point by any significant amount. The freezing points of both urine and ocular fluids are approximately –1 °C, mostly due to the ion content (Dobbs and DeVries 1975a; Turner et al. 1985; Ahlgren et al. 1988). These two fluids are therefore supercooled by 0.9 °C and would freeze in the presence of ice. In the Antarctic fish, urine does not freeze because the urethra, which is the only route of inward ice propagation, is normally closed by a strong muscular sphincter with a substantial amount of mucus. The ocular fluid cavity, on the other hand, does not communicate with the environment, and therefore the only way for ice entry is across the integument. The head skin surrounding the ocular orbit extends over the cornea as a transparent tissue and has been shown to act as a physical barrier to ice entry (Turner et al. 1985).

5 Noncolligative Lowering of the Freezing Point Through Adsorption-Inhibition

Impurities adsorbing to crystals are known to inhibit the growth of small crystals (see Raymond 1976). In many cases, impurities adsorb to and inhibit growth on crystal faces, and cause a marked change in the crystal habit (Buckley 1952; Butchard and Whetstone 1949). The inhibition of ice crystal growth below the equilibrium freezing point of water by the AFGPs and AFPs appears to be another example of the adsorption-inhibition phenomenon.

5.1 Adsorption

When solutions of AFGPs and AFPs are being frozen, the antifreeze molecules are preferentially retained in the ice, unlike other molecules of similar size and shape which are excluded into the unfrozen solution (Duman and DeVries 1972; Tomimatsu et al. 1976; Raymond and DeVries 1977). The incorporation of antifreeze into the solid phase was attributed to antifreeze adsorption to ice (Raymond and DeVries 1977).

The details of the mechanism of adsorption to ice by the antifreezes are not completely understood. However, it is likely that hydrogen bonding must be involved, because all of the hydroxyl, carboxyl, and amino groups of the antifreezes can potentially form hydrogen bonds with the oxygens or hydrogens in the ice lattice. This is supported by the fact that chemical modifications of almost any of the hydroxyls of the carbohydrate moiety of AFGPs (Duman and DeVries 1972; Shier et al. 1972; Lin et al. 1972), or of the carboxyl groups of the aspartic and glutamic acid residues in some AFPs (Duman and DeVries 1976), or of the amino groups of the lysines in the sculpin AFP (Schrag and DeVries, unpublished), all lead to loss of the antifreeze effect.

It has been proposed that adsorption involves a lattice match between the hydrogen-bonding groups in the antifreeze molecules and the oxygen atoms in the ice lattice (DeVries and Lin 1977b; DeVries 1984). Examination of space-filling models of AFGPs reveals that many of the hydroxyls of the disaccharide side chain are separated from each other by 4.5 Å, the same as the repeat spacing of water molecules in the ice lattice parallel to the a-axes. In addition, assuming a completely extended conformation of the AFGP molecule, the alternate carbonyl groups on the peptide backbone are 7.3 Å apart, which matches the repeat spacing (7.36 Å) along the c-axis in the ice lattice (Fletcher 1970).

The AFPs of winter flounder, Alaskan plaice, and short-horn sculpin contain clusters of polar amino acids separated by long sequences of nonpolar alanine residues (Fig. 2). The polar clusters usually contain threonine and aspartate separated by two alanines. Circular dichroism studies of these peptides indicate that they are alpha-helices (Raymond et al. 1977; Ananthanarayanan and Hew 1977; Hew et al. 1985). In such a conformation, the AFP molecule is amphiphilic, with the nonpolar and polar residues positioned on opposite sides of the helix. Measurements of the distance between the polar side chains of the aspartate and threonine in such a conformation show they are separated by 4.5 Å, matching the repeat spacing in the ice lattice parallel to the a-axes. This lattice match again suggests that the peptides orient themselves in a specific pattern on the ice lattice and bind to it by means of hydrogen bonding. Using X-ray crystallographic techniques, Yang et al. (1988) confirmed the helical structure of the flounder AFP, but proposed that dipole-dipole interactions between the AFP and ice, rather than a lattice match, align the AFPs on the prism faces prior to binding.

Knight and DeVries (1988) recently determined the ice crystal plane where adsorption occurs for a number of antifreezes. An oriented ice single crystal was allowed to grow slowly in very dilute antifreeze solution into an ice single-crystal hemisphere. After etching by sublimation of the ice hemisphere surface, the regions where antifreeze molecules have been incorporated, often oval in shape, take on the appearance of finely ground glass, while the rest of the surface remains mirror smooth. The tangent to the center of these etched regions is the adsorption plane whose orientation was determined with a universal stage. Using this method, it was found that different antifreezes have different preferential planes of adsorption (Fig. 4). The AFPs from the Antarctic eel pouts, *Rhigophila dearborni* and *Austrolycichthys brachycephalus*, adsorb to the primary prism planes $\{10\bar{1}0\}$; AFGP1–5 from the Antarctic cod, *Dissostichus mawsoni*, adsorb to a high index plane $(72\bar{9}0)$ very close to the primary prism plane; the short-horn sculpin AFP adsorbs to the secondary prism planes $\{2\bar{1}\bar{1}0\}$; the winter flounder and Alaskan plaice AFPs adsorb to the pyramidal planes $\{20\bar{2}1\}$, instead of the originally proposed primary prism planes, $\{10\bar{1}0\}$; and the sea raven AFP adsorbs to another pyramidal plane $\{11\bar{2}1\}$. None of the antifreezes adsorb on or near the basal plane (Knight and DeVries 1988). The molecular alignments of the flounder, plaice, and sculpin AFPs were all found to be along the $<0\bar{1}\bar{1}2>$ direction, which is found in both the $\{20\bar{2}1\}$ and $\{2\bar{1}\bar{1}0\}$ planes, and which has a

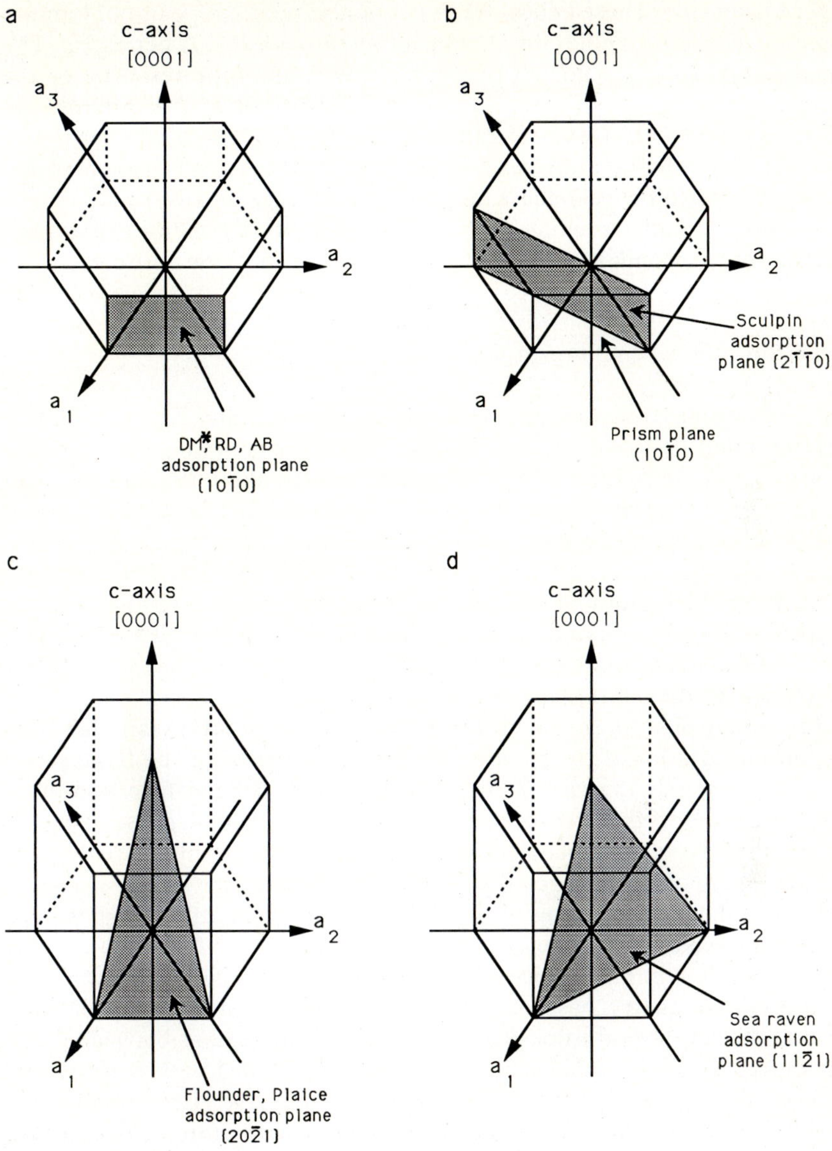

Fig. 4a-d. Plane of adsorption of antifreeze from (**a**) *DM* Dissostichus mawsoni*. **a** Mixture of antifreeze glycopeptides, AFGP1–5, adsorbs to a high index plane (72$\bar{9}$0) that is very close to the primary prism plane (10$\bar{1}$0). For ease of illustration, it is drawn as (10$\bar{1}$0). *RD Rhigophila dearborni*, *AB Austrolycichthys brachycephalus*. **b** Sculpin; **c** winter flounder, Alaskan plaice; **d** sea raven

16.7 Å repeat spacing. These helical AFPs contain sequence repeats of 11 amino acids, ending with a threonine (Fig. 2) which constitutes a 16.5 Å per repeat. The rotational freedom of the polar side chain of the threonine residues, or the flexibility in the helix itself, might readily compensate for the 0.2 Å difference to allow for a lattice match (DeVries, unpublished).

The fact that different antifreezes have specific planes of adsorption argues for a structural match between these molecules and the ice for adsorption to take place. A complete understanding of the mechanism of adsorption will require elucidation of the secondary and/or tertiary structures of these antifreezes.

5.2 Inhibition

Inhibition of crystal growth by adsorbed impurities is a rather common phenomenon (Raymond 1976; Addadi and Weiner 1985; Berman et al. 1988). Growth inhibition is often accompanied by changes in crystal habit. The crystal faces where adsorbents bind have retarded growth, and therefore become expressed when growth layers are deposited on adjacent faster-growing faces. Such a phenomenon has been observed in ice single crystals in the presence of antifreeze (Raymond et al. 1989). Both AFGPs and AFPs cause ice single crystals to assume unusual and strikingly similar habits. At temperatures within the hysteresis gap, AFGP1–5 and AFPs completely inhibit growth on the prism faces, but allow limited growth on the basal plane. As growth layers are laid down on the basal plane, hexagonal pits develop within the basal plane, while pyramidal faces develop on the exterior of the ice crystal. Growth stops when the basal plane becomes fully pitted. In the case of some AFPs, growth on the pit faces continues (the pit faces are stepped or irregular, and therefore may contain sites having the same orientation as the basal plane to allow continued growth) until the ice crystal assumes the shape of a bihexagonal pyramid (Raymond et al. 1989).

The noncolligative lowering of the freezing point by antifreezes has been proposed to be based on an increase in surface free energy following adsorption-inhibition. Assuming that ice crystal growth occurs via water molecules joining the crystal on the basal planes at the steps, the antifreeze molecules adsorbed at the step would force growth to occur in the regions between them. Consequently, the step would be divided into many small fronts separated by adsorbed antifreeze molecules (Fig. 5). These fronts will have a large surface area compared to their volume because of their increased curvature, which in turn raises the surface free energy. Growth between the molecules will stop when the ratio of surface area to volume exceeds a critical point. Greater supercooling will be required to remove the energy from the system to allow the small fronts to propagate, i.e., for freezing to continue. In other words, the freezing point of water is lowered. The spacing between the adsorbed antifreeze molecules appears to be a function of their concentration, size, and shape. If certain assumptions are made about the density and randomness of the antifreeze molecules on the crystal face, it can be stated that the amount of supercooling (or freezing point depression) is proportional to the square root of the concentration. Using this relationship, it can

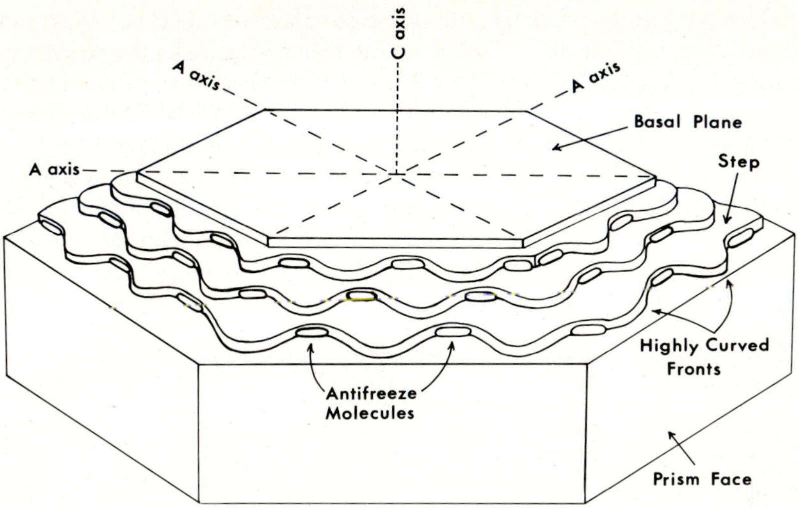

Fig. 5. Model illustrating adsorption-inhibition as a mechanism of freezing point depression of water. Adsorbed antifreeze molecules on the basal plane stop growth steps because they divide long straight fronts into many small fronts. These small fronts become highly curved and cause an increase in the surface free energy of the ice crystal. In order for growth to occur the temperature must be lowered

be shown that there is good agreement between freezing point depression curves obtained experimentally and those derived from the above relationship (Raymond and DeVries 1977).

6 Adsorption-Inhibition and Freezing Avoidance

The mechanism of adsorption-inhibition for the freezing avoidance of fish implies that ice is present somewhere in the live fish. Our recent studies show indirect evidence for its presence. Fish taken from the ice-laden surface waters and subjected to low temperatures (–2.7 °C) in ice-free seawater will freeze with ice visibly propagating through their superficial tissues as well as through the clear fluids of their eye, and death occurs within 1 min. Freezing occurs in these fish even if they have been held for several hours at –1.2 °C, a temperature that would ensure melting of any external ice crystals associated with their gills or integument (DeVries, unpublished). These observations imply that they harbor ice crystals somewhere within their bodies. However the ice can be "melted" by holding the fish for several minutes at 0 °C. The fish can then be supercooled to at least –7 °C in ice-free seawater without freezing and with no apparent ill effect. Subsequent exposure to ice-laden seawater at –1.9 °C for 1 h "reinoculates" the fish and they again freeze at –2.7 °C in ice-free water. Blood from these fishes taken directly from their natural environment does not freeze spontaneously at –2.7 °C. This suggests that ice is present at sites other than the circulatory system. A logical site of endogenous ice is the intestinal fluid since the fish drinks ice-laden seawater; unless the ice crystals in the intestinal fluid can exit with other excrement, they must exist in the fish for its lifetime, because they can never melt at the ambient temperature at which the fish lives. The melting point of ice in the body fluids of Antarctic fish is approximately −1 °C, a temperature well above the mean annual ambient temperature of −1.86 °C. However, we have recently observed that the cryopelagic Antarctic fish, *Pagothenia borchgrevinki*, make excursions down to 200–300 m depths for feeding, where the effect of hydrostatic pressure on the melting point of ice should melt the external ice associated with the fish.

Acknowledgment. This research is supported by NSF grant DPP87-16296 to ALD.

References

Addadi L, Weiner S (1985) Interactions between acidic protein and crystals: stereochemical requirements in biomineralization. Proc Natl Acad Sci USA 882:4110–4114

Ahlgren JA, DeVries AL (1984) Comparison of antifreeze glycopeptides from several antarctic fishes. Polar Biol 3:3–97

Ahlgren JA, Cheng CC, Schrag JD, DeVries AL (1988) Freezing avoidance and the distribution of antifreeze glycopeptides in body fluids and tissues of Antarctic fish. J Exp Biol 137:549–563

Ananthanarayanan VS, Hew CL (1977) Structural studies on the freezing point-depressing protein of the winter flounder *Pseudopleuronectes americanus*. Biochem Biophys Res Commun 74:685–689

Ananthanarayanan VS, Slaughter D, Hew CL (1986) Antifreeze proteins from the ocean pout, *Macrozoarces americanus*: circular diochroism spectral studies on the native and denatured states. Biochim Biophys Acta 870:154–159

Andriashev AP (1970) Cryopelagic fishes in the Arctic and Antarctic and their significance in polar ecosystems. In: Holdgate MW (ed) Antarctic ecology, vol 1. Academic Press, London, p 297

Berman A, Addadi L, Weiner S (1988) Interactions of sea-urchin skeleton macromolecules with growing calcite crystal – a study of intracrystalline proteins. Nature 331:546

Black VS (1951) Some aspects of the physiology of fish. II. Osmotic regulation in teleost fishes. Univ Toronto Stud Biol Ser 59, 71:53–89

Boyd RB, DeVries AL (1983) The seasonal distribution of anionic binding sites in the basement membrane of the kidney glomerulus of the winter flounder *Pseudopleuronectes americanus*. Cell Tissue Res 234:271–277

Buckley HE (1952) Crystal growth. Wiley, New York, pp 339

Butchard A, Whetstone J (1949) The effect of dyes on crystal habits of some oxy-salts. Discuss Faraday Soc 5:254–261

Cheng CC, DeVries AL (1989) Structures of antifreeze peptides from the Antarctic eel pout, *Austrolycichthys brachycephalus*. Biochim Biophys Acta 997:55–64

DeVries AL (1970) Freezing resistance in Antarctic fishes. In: Holdgate MW (ed) Antarctic ecology, vol I. Academic Press, London, pp 320

DeVries AL (1971) Glycoproteins as biological antifreeze agents in Antarctic fishes. Science 172:152–1155

DeVries AL (1974) Survival at freezing temperatures. In: Sargent JS, Mallins DW (eds) Biochemical and biophysical perspectives in marine biology, vol I. Academic Press, London, pp 289

DeVries AL (1980) Biological antifreezes and survival in freezing environments. In: Gilles R (ed) Animals and environmental fitness. Pergamon, Oxford, pp 583

DeVries AL (1982) Biological antifreeze agents in coldwater fishes. Comp Biochem Physiol A73:627–640

DeVries AL (1984) Role of glycopeptides and peptides in inhibition of crystallization of water in polar fishes. Philos Trans R Soc Lond B304:575–588

DeVries AL (1986) Antifreeze glycopeptides and peptides: interactions with ice and water. Methods Enzymol 127:293–303

DeVries AL, Lin Y (1977a) The role of glycoprotein antifreezes in the survival of Antarctic fishes. In: Llano GA (ed) Adaptations within Antarctic ecosystems. Gulf, Houston, Texas, pp 439

DeVries AL, Lin Y (1977b) Structure of a peptide antifreeze and mechanism of adsorption to ice. Biochim Biophys Acta 495:88–392

DeVries AL, Wohlschlag DE (1969) Freezing resistance in some antarctic fishes. Science 163:1074–1075

DeVries AL, Komatsu SK, Feeney RE (1970) Chemical and physical properties of freezing point-depression glycoproteins from Antarctic fishes. J Biol Chem 245:2901–2913

DeVries AL, Vandenheede J, Feeney RE (1971) Primary structure of freezing point-depressing glycoproteins. J Biol Chem 246:305–308

Dobbs GH, DeVries AL (1975a) Renal function in antarctic teleost fishes: serum and urine composition. Mar Biol 29:59–70

Dobbs GH, DeVries AL (1975b) Aglomerular nephron of Antarctic teleosts: a light electron microscopic study. Tissue Cell 7:159–170

Dobbs GH, Lin Y, DeVries AL (1974) Aglomerularism in Antarctic fish. Science 185:793–794

Duman JG, DeVries AL (1972) Freezing behavior of aqueous solutions of glycoproteins from the blood of antarctic fish. Cryobiology 9:469–472

Duman JG, DeVries AL (1975) The role of macromolecular antifreezes in cold water fishes. Comp Biochem Physiol 52A:93–199

Duman JG, DeVries AL (1976) Isolation, characterization and physical properties of protein antifreezes from the winter flounder, *Pseudopleuronectes americanus*. Comp Biochem Physiol 53b:375–380

Eastman JT, DeVries AL (1986) Renal glomerular evolution in Antarctic notothenioid fishes. J Fish Biol 29:649–662

Eastman JT, DeVries AL, Coalson RE, Nordquist RE, Boyd RB (1979) Renal conservation of antifreeze peptide in antarctic eel pout, *Rhigophila dearborni*. Nature 282:217–218

Feeney RE, Yeh Y (1978) Antifreeze proteins from fish bloods. Adv Protein Chem 32:191–282

Fletcher GL (1977) Circannual cycles of blood plasma freezing point and Na^+ and Cl^- concentrations in Newfoundland winter flounder (*Pseudopleuronectes americanus*) correlation with water temperature and photoperiod. Can J Zool 55:789–795

Fletcher NH (1970) The chemical physics of ice. Cambridge University Press, Cambridge, pp 111

Gordon MS, Amdur BH, Scholander PF (1962) Freezing resistance in some northern fishes. Biol Bull Mar Biol Lab, Woods Hole 122:52–62

Gourlie B, Lin Y, Powers D, DeVries AL, Huang RC (1984) Winter flounder antifreeze protein: evidence for a multigene family. J Biol Chem 259:14960–14965

Hargens AR (1972) Freezing resistance in polar fishes. Science 176:184–186

Hew CL, Joshi S, Wang NC, Kao MH, Ananthanarayanan VS (1985) Structures of shorthorn sculpin antifreeze polypeptides. Eur J Biochem 151:167–172

Hudson AP, DeVries AL, Haschemeyer AEV (1979) Antifreeze glycoprotein biosynthesis in Antarctic fishes. Comp Biochem Physiol 62B:179–183

Karnaky KJ (1986) Structure and function of the chloride cell of *Fundulus heteroclitus* and other teleosts. Am Zool 26:209–224

Knight CA, DeVries AL (1988) The prevention of ice crystal growth from water by "antifreeze proteins". In: Wagner PE, Valli G (eds) Atmospheric aerosol and nucleation. Springer, Berlin Heidelberg New York Tokyo, pp 717

Komatsu SK, DeVries AL, Feeney RE (1970) Studies of the structure of the freezing point-depressing glycoproteins from an antarctic fish. J Biol Chem 245:2901–2908

Li XM, Trinh KY, Hew CL, Buettner B, Baenziger J, Davies PL (1985) Structure of an antifreeze polypeptide and its precursor from the ocean pout, *Macrozoarces americanus*. J Biol Chem 260:12904–12909

Lin Y, Duman JG, DeVries AL (1972) Studies on the structure and activity of low molecular weight glycoproteins from an antarctic fish. Biochim Biophys Res Commun 46:87–92

Littlepage JL (1965) Oceanographic investigations in McMurdo Sound, Antarctica. In: Llano GA (ed) Antarctic research series, vol 5, Biology of antarctic seas II, American Geophysical Union, Washington, DC, p 1

Morris HR, Thompson MR, Osuga DT, Ahmed AI, Chan SM, Vandenheede JR, Feeney RE (1978) Antifreeze glycoproteins from the blood of an Antarctic fish. J Biol Chem 253:5155–5162

Ng N, Trinh YK, Hew CL (1986) Structure of an antifreeze polypeptide precursor from the sea raven, *Hemitripterus americanus*. J Biol Chem 261:15690–15696

O'Grady SM, DeVries AL (1982) Osmotic and ionic regulation in polar fishes. J Exp Mar Biol Ecol 57:219–228

O'Grady SM, Clarke A, DeVries AL (1982a) Characterization of glycoprotein antifreeze biosynthesis in isolated hepatocytes from *Pagothenia borchgrevinki*. J Exp Zool 220:179–189

O'Grady SM, Ellory JC, DeVries AL (1982b) Protein and glycoprotein antifreezes in the intestinal fluid of polar fishes. J Exp Biol 98:429–438

O'Grady SM, Ellory JC, DeVries AL (1983) The role of low molecular weight antifreeze glycopeptides in the bile and intestinal fluid of Antarctic fishes. J Exp Biol 104:149–162

Osuga DT, Feeney RE (1978) Antifreeze glycoproteins from Arctic fish. J Biol Chem 253:5338–5343

Petzel D, Reisman H, DeVries AL (1980) Seasonal variation of antifreeze peptide in the winter flounder, *Pseudopleuronectes americanus*. J Exp Zool 211:63–69

Potts WTW, Parry G (1964) Osmotic and ionic regulation in animals, vol 19. Pergamon, Oxford, pp 171

Raymond JA (1976) Adsorption inhibition as a mechanism of freezing resistance in polar fishes. PhD Thesis, University of California, San Diego

Raymond JA, DeVries AL (1977) Adsorption inhibition as a mechanism of freezing resistance in polar fishes. Proc Natl Acad Sci USA 74:2589–2593

Raymond JA, Lin Y, DeVries AL (1975) Glycoproteins and protein antifreeze in two Alaskan fishes. J Exp Zool 193:25–130

Raymond JA, Radding W, DeVries AL (1977) Circular dichroism of protein and glycoprotein fish antifreeze. Biopolymers 16:2575–2578

Raymond JA, Wilson P, DeVries AL (1989) Inhibition of growth of non-basal planes in ice by fish antifreezes. Proc Natl Acad Sci USA 86:881–885

Schneppenheim R, Theede H (1982) Freezing point depressing peptides and glycopeptides from arctic-boreal and antarctic fish. Polar Biol 1:115–123

Scholander PF, Vandam L, Kanwisher JW, Hammel HT, Gordon MS (1957) Supercooling and osmoregulation in arctic fish. J Cell Comp Physiol 49:5–24

Schrag JD, Cheng CC, Panico M, Morris HR, DeVries AL (1987) Primary and secondary structure of antifreeze peptides from arctic and antarctic zoarcid fishes. Biochim Biophys Acta 915:357–370

Shier WT, Lin Y, DeVries AL (1972) Structure and mode of action of glycoproteins from an antarctic fish. Biochim Biophys Acta 263:406–413

Shier WT, Lin Y, DeVries AL (1975) Structure of the carbohydrate of antifreeze glycoproteins from an antarctic fish. FEBS Lett 54:135–138

Slaughter D, Fletcher GL, Ananthanarayanan VS, Hew CL (1981) Antifreeze proteins from the sea raven, *Hemitripterus americanus*. J Biol Chem 256:2022–2026

Tomimatsu Y, Scherer J, Yeh Y, Feeney RE (1976) Raman spectra of a solid antifreeze glycoprotein and its liquid and frozen aqueous solutions. J Biol Chem 251:2290–2298

Turner JD, Schrag JD, DeVries AL (1985) Ocular freezing avoidance in antarctic fish. J Exp Biol 118:121–131

Van Voorhies WV, Raymond JA, DeVries AL (1978) Glycoproteins as biological antifreeze agents in the cod *Gadus ogac* (Richardson). Physiol Zool 51:347–353

Yang DSC, Sax M, Chakrabartty A, Hew CL (1988) Crystal structure of an antifreeze polypeptide and its mechanistic implications. Nature 333:232–237

An Overview of the Molecular Structure and Functional Properties of the Hemoglobins of a Cold-Adapted Antarctic Teleost

R. D'Avino[1], C. Caruso[1], L. Camardella[1], M.E. Schininà[2], B. Rutigliano[1], M. Romano[1], V. Carratore[1], D. Barra[2], and G. di Prisco[1]

1 Introduction

During the Paleozoic and Mesozoic eras, over 300 Ma, Antarctica, together with South America, Africa, India, Australia, and New Zealand, was part of the supercontinent of Gondwana. It enjoyed a much warmer climate, with forests and warm-weather animals, than that of the dry, bitterly cold desert of current times. Gondwana began to break up approximately in 135 Ma, near the Jurassic-Cretaceous boundary; the final separation of the fragments occurred in the early Tertiary. The continental drift carried Antarctica close to the present position at the beginning of the Cenozoic era, about 65 Ma (King 1958; Wilson 1963; Adie 1970; Craddock 1970; Dietz and Sproll 1970; Frakes and Crowell 1970; Schopf 1970; Kennett 1977). Progressive cooling followed, with glaciation and ice sheet formation possibly beginning as early as 40 Ma (Tanner 1968; Denton et al. 1970; Le Masurier 1970). The isolation of Antarctica became complete 25–22 Ma, near the Oligocene-Miocene boundary, with the opening of the Drake Passage (Barker and Burrell 1977; Kennett 1977). This event permitted the onset of the Circum-Antarctic Current and the resulting development of the Antarctic Convergence: a well-defined, roughly circular, oceanic frontal system, currently running between 50° S and 60° S, where the surface layers of the north-flowing Antarctic waters sink beneath the less cold, less dense sub-Antarctic waters. The cooling of the environment steadily proceeded, reaching the present, extreme climatic conditions. For this reason, at the Convergence, far more north than the coastline of Antarctica, the ocean temperature in winter is 1–2 °C (in the summer it reaches 4–5 °C, but only in the 50-m surface layer); just north of the Convergence, there is an abrupt increase of 3 °C. For the Antarctic marine fauna, the Convergence appears to be a factor of the utmost importance.

Throughout the year, the temperature of the coastal Antarctic waters, in which fish from temperate waters would rapidly freeze and could not survive, is −1.87 °C, the equilibrium temperature of the ice-salt water mixture. Nevertheless, the oxygen-rich Antarctic waters support a wealth of marine life. During the increasing geographic and climatic isolation south of the Convergence, the physiology of Antarctic teleosts became gradually adjusted to tolerate the pro-

[1] Institute of Protein Biochemistry and Enzymology, C.N.R., Via Marconi 10, 80125 Naples, Italy.
[2] Department of Biochemical Sciences and C.N.R. Center of Molecular Biology, University "La Sapienza", Rome, Italy.

Guido di Prisco (Ed)
Life Under Extreme Conditions

gressive cooling. Being exposed to unique ambient conditions, fish developed cold adaptation (Wohlschlag 1964), the evolutionary counterpart of cold acclimation; in fact, their exposure to water temperatures of a few degrees above zero has lethal effects (Somero and DeVries 1967). The Convergence became a natural barrier to migration in both directions; it represents, therefore, a key factor for fish isolation and evolution.

In the process of cold adaptation, the evolutionary trends of Antarctic teleosts include unique specializations, some of which are illustrated and discussed in other contributions to this volume. One of these specializations is the modification of the hematological characteristics.

The blood of these fish has acquired some features which clearly differentiate them from fish of temperate or tropical climates. It contains fewer erythrocytes and less hemoglobin (Everson and Ralph 1968; Hureau et al. 1977; Wells et al. 1980) in comparison with non-Antarctic fish (Coburn and Fischer 1973; Larsson et al. 1976): the erytrocyte number and the hemoglobin concentration can be as low as one-tenth and one-fifth, respectively. At the extreme end of such evolution, the colorless blood of the 16 species of the family Channichthyidae (the only vertebrates which display this feature) totally lacks hemoglobin (Ruud 1954). Hureau et al. (1977) have reported the presence of a very small number of erythrocyte-like cells; the physiological relevance of the blood cells of Channichthyidae may be traced in finding that each one of these cells contains much more activity of glucose-6-phosphate dehydrogenase, an enzyme of great metabolic significance, than a cell of red-blooded Antarctic fish (di Prisco and D'Avino 1989).

At the temperature of the environment, the solubility of oxygen in sea water is increased; the affinity of hemoglobin for oxygen also becomes higher, since the binding of the gas is usually an exothermal process. On the other hand, the sub zero seawater temperature would induce a great increase in the viscosity of a corpusculate fluid such as blood. The decrease in the number of erythrocytes and in the amount of hemoglobin, counteracting the temperature-induced viscosity increase, greatly facilitates the cardiac work. It brings the energy demand to levels tolerable to the organism and may therefore be considered among the components of cold adaptation.

The biochemistry and physiology of respiration are therefore confronted with rather exceptional conditions. The number of intriguing questions that arises justifies the initiation of a thorough investigation on the relationship between the molecular structure and the oxygen-binding properties of hemoglobins isolated from the blood of several species of Antarctic teleosts. The aim of this study is to gain an insight into the molecular basis of fish cold adaptation, as well as into the evolutionary history of the development of these features in fish, during the isolation that followed the separation and drift of Antarctica from Gondwana.

Notothenia coriiceps neglecta is one of the species biologically adjusted to survive the environmental temperature. The erythrocytes of this species contain two hemoglobins, which have been purified by ion-exchange chromatography and obtained in crystalline form: Hb 1 (about 90% of total) and Hb 2 (about 5%; D'Avino and di Prisco 1989).

This chapter reports the complete amino acid sequence of the two hemoglobins of *N. coriiceps neglecta*, as well as the effect of pH and of some physiological effectors (organic phosphates) on the oxygen-binding properties of erythrocytes and hemoglobins (Bohr and Root effects).

2 Methodology

Materials

Specimens of *N. coriiceps neglecta* (genus, *Notothenia*; family, Nototheniidae) were collected as described (D'Avino and di Prisco 1989).

L-1-tosylamide-2-phenylethylchloromethylketone-trypsin and chymotrypsin were from Worthington Biochemical Co.; carboxypeptidases A and B from Sigma Chemical Co.; *Staphylococcus aureus* V8 endoproteinase Glu-C from Boehringer Mannheim; dithiothreitol, 4-*N,N*-dimethylaminoazobenzene 4′-isothiocyanate and phenylisothiocyanate from Pierce Chemical Co.; Sequanal-grade trifluoroacetic acid from Fluka AG; HPLC-grade acetonitrile from Carlo Erba; Sequencer reagents from Applied Biosystems. All other reagents were of the highest purity commercially available.

Hemoglobin and Globin Purification and Carboxymethylation

Hb 1, Hb 2, and their globin chains were purified by ion-exchange chromatography and by RP-HPLC, as previously described (D'Avino and di Prisco 1989).

S-carboxymethylated globins were digested with trypsin and chymotrypsin (D'Avino et al. 1989); Hb 2 α-chains were cleaved by CNBr (Gross and Witkop 1961) and some of the CNBr fragments further digested with endoproteinase Glu-C (Drapeau 1977). Deacylation of the amino terminus of chains was achieved as described (D'Avino et al. 1989). Cleavage of the Asp-Pro bond was performed according to Schinina' et al. (1988).

Peptide Purification; Amino Acid Analysis

Peptides mixtures from proteolytic and chemical cleavages were separated by RP-HPLC; amino acid analyses were performed as described (D'Avino et al. 1989).

Amino Acid Sequence Analysis

Amino acid sequence analysis was performed by manual as well as automated Edman degradation, using protein-peptide sequencing systems, models 470 and 477A, from Applied Biosystems, equipped with a 120A analyser for the on-line detection of PTH-amino acids.

Peptide Nomenclature

Peptides were named according to the enzyme or the chemical reagent used in proteolysis: T (trypsin), CB (CNBr).

Oxygen Binding

The oxygen saturation was measured calculating the average of the absorbance difference at three wavelengths (540, 560, and 575 nm) between the spectra observed at a given pH before and after addition of a few crystals of sodium dithionite, which caused complete deoxygenation of hemoglobin. Oxygen equilibrium curves were obtained tonometrically as described (Giardina and Amiconi 1981). Buffers were 100 mM Tris-HCl, pH 8.5–7, and Bistris-HCl, pH 7–5.5. Endogenous organic phosphates were "stripped" by running the hemolysate through a small column of a mixed-bed ion-exchange resin, Dowex AG 501 X8 (D).

3 Results

The HPLC elution time, electrophoretic mobility, molecular weight, and amino acid composition of the globin chains indicated that Hb 1 and Hb 2 have the β-chain in common and differ by the α-chain (D'Avino and di Prisco 1989).

3.1 Hb 1 α-Chain

The amino acid sequence of the α-chain of the main component Hb 1, elucidated by repetitive, modified manual Edman degradation, has been recently described (D'Avino et al. 1989). It was the first report of the sequence of a globin from a marine poikilotherm and, moreover, the first one referring to a cold-adapted Antarctic teleost.

3.2 β-Chain, in Common to Hb 1 and Hb 2

Figure 1 illustrates the complete amino acid sequence, essentially established by automated Edman degradation, of the β-chain of the two hemoglobins present in the erythrocytes of *N. coriiceps neglecta*.

Following tryptic digestion of the carboxymethylated globin, the peptides were fractionated by RP-HPLC on a μBondapak C_{18} column according to a described procedure (D'Avino et al. 1989) and were obtained in pure form in a single chromatographic run.

Starting from the N-terminus, the sequence of 47 residues was elucidated by automated sequencing of the intact protein. Similarly to other fish hemoglobin β-chains, Val was identified at the amino terminus. Tryptic peptides were

10 20 30
Val-Asn-Trp-Ser-Asp-Ser-Glu-Arg-Ala-Ile-Ile-Thr-Asp-Ile-Phe-Ser-His-Met-Asp-Tyr-Asp-Asp-Ile-Gly-Pro-Lys-Ala-Leu-Ser-Arg-
T1 T2 T3

40 50 60
-Cys-Leu-Ile-Val-Tyr-Pro-Trp-Thr-Gln-Arg-His-Phe-Ser-Gly-Phe-Gly-Asn-Leu-Tyr-Asn-Ala-Glu-Ala-Ile-Leu-Gly-Asn-Ala-Asn-Val-
T4 T5

70 80 90
-Ala-Ala-His-Gly-Ile-Lys-Val-Leu-His-Gly-Leu-Asp-Arg-Gly-Val-Lys-Asn-Met-Asp-Lys-Ile-Val-Asp-Ala-Tyr-Ala-Glu-Leu-Ser-Met-
T6 T7 T8 T9

100 110 120
-Leu-His-Ser-Glu-Lys-Leu-His-Val-Asp-Pro-Asp-Asn-Phe-Lys-Leu-Leu-Ser-Asp-Cys-Ile-Thr-Ile-Val-Val-Ala-Ala-Lys-Met-Gly-His-
T10 T11

130 140
-Ala-Phe-Thr-Pro-Glu-Ile-Gln-Gly-Ala-Phe-Gln-Lys-Phe-Leu-Ala-Val-Val-Val-Ser-Ala-Leu-Gly-Lys-Gln-Tyr-HisCOOH
T12 T13 T14

Fig. 1. Complete amino acid sequence of the β-chain of *N. coriiceps neglecta* Hb 1 and Hb 2. Sequence portions established by automated Edman degradation and carboxypeptidase-A digestion, and proteolytically generated peptides used to establish the complete sequence, are indicated

sequenced to position 104 by manual and automated degradation; their alignment was unequivocally established by analogy with the corresponding sequences of β-chains of hemoglobin from other fish (Grujić-Injac et al. 1980; Braunitzer and Rodewald 1980; Barra et al. 1983; Petruzzelli et al. 1984; Rodewald et al. 1987). The cleavage of the peptide bond between Asp-99 and Pro-100 (Schinina' et al. 1988), followed by reaction with *o*-phthalaldehyde (Brauer et al. 1984) before sequencing, allowed the determination of the sequence between position 100 and 134 by automated Edman degradation. A manually sequenced, overlapping tryptic peptide (T13) extended the sequence to position 143. The C-terminal sequence of the globin chain, determined by carboxypeptidase A digestion and by manual sequence of a tryptic peptide (T14), is -Gln-Tyr-His.

The calculated molecular mass of this globin is 16 166 (the value of 17 000 had been calculated with SDS-PAGE); the sequence-deduced amino acid composition is in good agreement with that determined by total acid hydrolysis (D'Avino and di Prisco 1989).

The sequences of the β-chain of hemoglobins from four non-Antarctic fish species, as well as of human hemoglobin, are shown for comparison in Fig. 2.

3.3 Hb 2 α-Chain

The complete amino acid sequence of the α-chain of Hb 2 (Fig. 3) was elucidated by automated Edman degradation of HPLC-purified peptides, formed after cleavage of the protein with CNBr. Treatment with 70% formic acid (Schinina' et al. 1988) cleaved the Asp-Pro bond at position 95–96; following reaction with *o*-phthalaldehyde (Brauer et al. 1984), sequencing proceeded up to residue 132. The peptides were essentially aligned by analogy with the sequences of α-chains of other fish hemoglobins (Hilse and Braunitzer 1968; Powers and Edmundson 1972; Bossa et al. 1978; Braunitzer and Rodewald 1980; Rodewald et al. 1987; D'Avino et al. 1989; Petruzzelli et al. 1989).

Similar to Hb 1 α-chain (D'Avino et al. 1989), the N-terminus was acetylated, as demonstrated by fast-atom-bombardment mass spectrometry. Following partial deacetylation, by treatment of peptide CB1 with 30% trifluoroacetic acid for 2 h at 55 °C, sequencing proceeded to residue 24. CB1 was then subdigested with endoproteinase Glu-C and a peptide corresponding to residues 9–32 was isolated and sequenced, yielding the sequence of the C-terminal portion of CB1.

The calculated molecular mass is 15 857, in good agreement with the value of 15 600, obtained by SDS-PAGE; the sequence-deduced amino acid composition is also in good agreement with that determined by total acid hydrolysis (D'Avino and di Prisco 1989).

Figure 4 illustrates the alignment of α-chain amino acid sequences of hemoglobins of five non-Antarctic species, of Hb 1 and Hb 2 of *N. coriiceps neglecta,* and of human hemoglobin.

The similarity among the α- and β-chains of Hb 1 and Hb 2 and the available sequences of teleosts living under totally different conditions (carp, *Cyprinus*

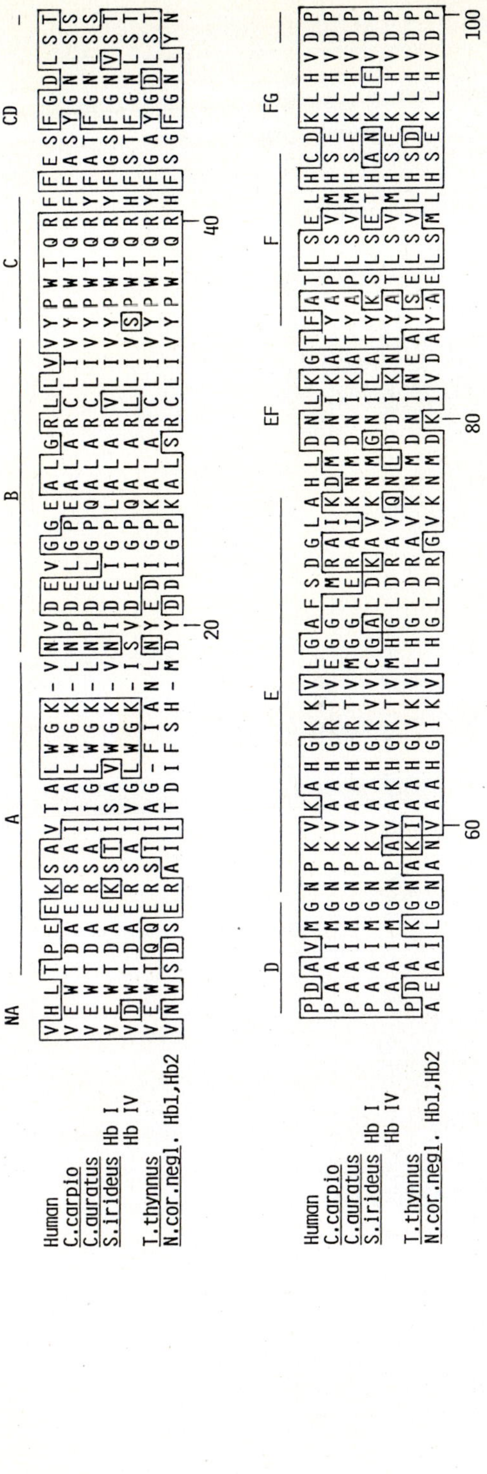

Fig. 2. Alignment of the amino acid sequence of the *β*-chain of *N. coriiceps neglecta* Hb 1 and Hb 2 with *β*-chain sequences of *C. carpio*, *C. auratus*, *S. irideus* Hb I and Hb IV, *T. thynnus*, and human. The one-letter amino acid notation is used. Residue identities in at least four sequences are *boxed*. The helical (*A, B, C, D, E, F, G, H*) and nonhelical (*NA, CD, EF, FG, GH, HC*) regions, as established for mammalian hemoglobins (Perutz 1969), are indicated above the human sequence

```
                                     10                                                   20                                                       30
Ac-Ser-Leu-Ser-Thr-Lys-Asp-Lys-Glu-Thr-Val-Lys-Ala-Phe-Trp-Ser-Lys-Val-Ser-Gly-Lys-Ser-Glu-Asp-Ile-Gly-Asn-Asp-Ala-Leu-Ser-
                                                            CB1

                                     40                                                   50                                                       60
-Arg-Met-Leu-Val-Val-Tyr-Pro-Gln-Thr-Lys-Thr-Tyr-Phe-Ser-His-Trp-Lys-Glu-Leu-Thr-Pro-Gly-Ser-Ala-Pro-Val-Arg-Lys-His-Gly-
                                                                   CB2

                                     70                                                   80                                                       90
-Met-Thr-Val-Met-Lys-Gly-Val-Gly-Asp-Ala-Val-Ser-Lys-Ile-Glu-Asp-Leu-Thr-Ala-Gly-Leu-Met-Glu-Leu-Ser-Glu-Leu-His-Ala-Phe-
          CB3                                       CB4

                                     100                                                  110                                                      120
-Thr-Leu-Arg-Val-Asp-Pro-Ala-Asn-Phe-Lys-Ile-Leu-Ser-His-Asn-Ile-Leu-Val-Val-Phe-Ala-Ile-Met-Phe-Pro-Lys-Glu-Phe-Thr-Ala-
                            CB5                                                                                                                   CB6

                                     130                                                  140
-Glu-Val-His-Val-Ser-Met-Asp-Lys-Phe-Leu-Ala-Ala-Leu-Ala-Arg-Ala-Leu-Ser-Glu-Lys-Tyr-ArgCOOH
                                                     CB7
```

Fig. 3. Complete amino acid sequence of the α-chain of *N. coriiceps neglecta* Hb 2. Sequence portions elucidated by automated Edman degradation and peptides generated by treatment with CNBr are indicated

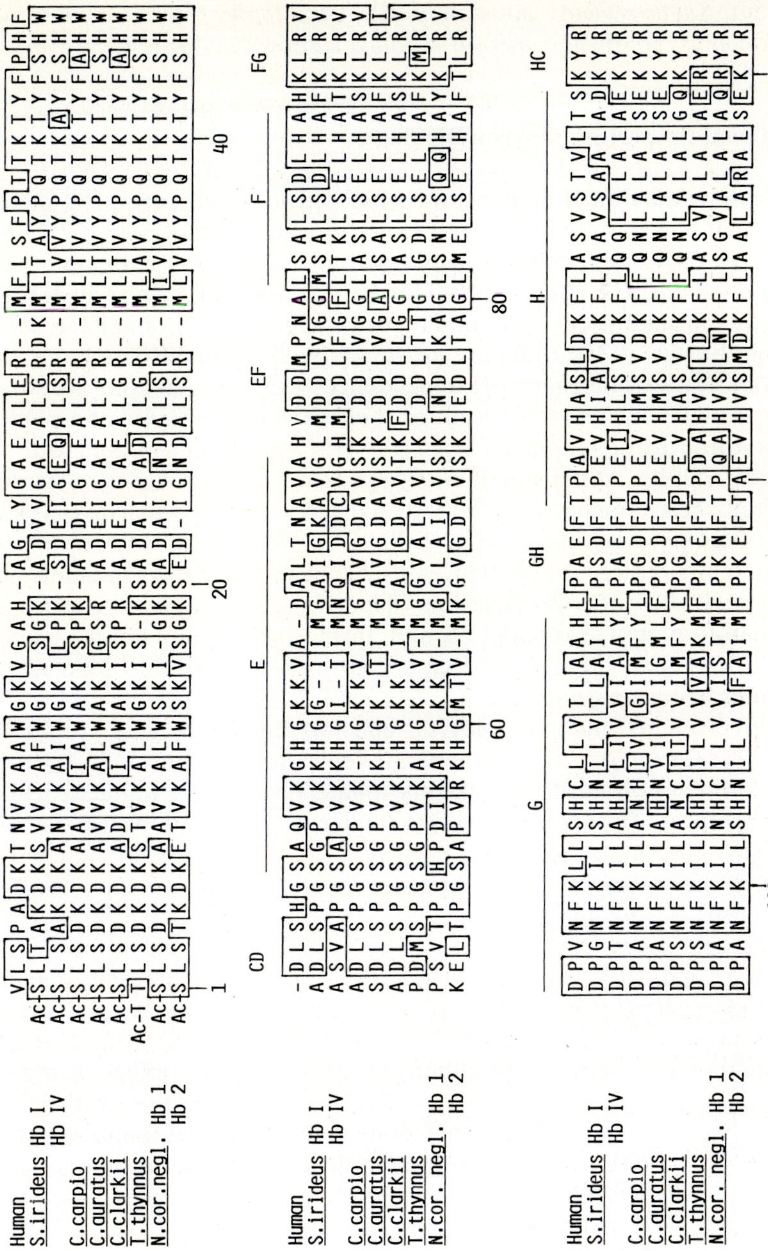

Fig. 4. Alignment of the amino acid sequence of the α-chains of *N. coriiceps neglecta* Hb 1 and Hb 2 with α-chain sequences of *C. carpio*, *C. auratus*, *C. clarkii*, *S. irideus* Hb I and Hb IV, *T. thynnus*, and human. Residue identities in at least five sequences are *boxed*. In α-chains, the helical region D is lacking

carpio; trout, *Salmo irideus*; catostomid fish, *Catostomus clarkii*; bluefin tuna, *Thunnus thynnus;* goldfish, *Carassius auratus*) is shown in Table 1.

3.4 Functional Properties: Oxygen Binding

A large, negative alkaline Bohr effect was observed when measuring the effect of pH on oxygen equilibrium curves in the erythrocytes, hemolysate (previously "stripped", in order to remove physiological effectors, such as organic phosphates), and purified Hb 1 and Hb 2 (di Prisco et al. 1988; D'Avino and di Prisco 1989). Wells and Jokumsen (1982) had reported appreciable Bohr effects shown by the hemolysates of five different species of Nototheniidae. The large Bohr effect was indicative of a sharp pH dependence of the oxygen affinities, which were also strongly sensitive to physiological effectors (Cl^-, organic phosphates).

The erythrocytes of *N. coriiceps neglecta* were found to bind oxygen with very low apparent affinity. At pH 7 and 20 °C, $p_{\frac{1}{2}}$ was 120 mmHg, compared with 25 mmHg of the well-characterized trout erythrocytes. However, when investigating the effect of temperature in the range 2–30 °C (D'Avino and di Prisco 1989), a decrease in temperature, as expected, caused an increase in oxygen affinity: for example, at pH 7.0, $p_{\frac{1}{2}}$ of erytrocytes was 70 mmHg at 2 °C. In Antarctic fish, the body temperature rarely exceeds 1 °C and, furthermore, the pH of the blood is 8.1–8.25 (Tetens et al. 1984); both factors contribute to raise the blood oxygen affinity under living conditions.

The effect of pH and physiological effectors on the oxygen equilibrium isotherms (20 °C) and on subunit cooperativity is shown in Fig. 5. At lower pH and in the presence of inositol hexakisphosphate, it was not possible to achieve saturation in air at atmospheric pressure; under these conditions, i.e., when $p_{\frac{1}{2}}$ values were higher than the partial pressure of oxygen in air, the latter were calculated by zero extrapolation of the experimental $\lg[\bar{Y}/(1-\bar{Y})]$ points of the Hill plots.

The Hill coefficient *n*, the other parameter of Hill's empyrical equation: $\bar{Y}/(1-\bar{Y}) = Kp^n$, indicated that the cooperativity of oxygen binding, maximal at the physiological pH values of 7.5–8 ($n_{\frac{1}{2}} = 3$) disappeared at pH 6, and even at pH 7 in the presence of organic phosphates ($n_{\frac{1}{2}} = 1$).

In experiments aimed at quantitatively measuring the effect of organic phosphates, the maximal enhancement of the Bohr effect was observed at 0.1–1 mM ATP; the concentration of the nucleotide in the blood was found to be high enough to saturate hemoglobin in the erythrocytes at nonalkaline pH values (D'Avino and di Prisco 1989).

The large reduction of oxygen affinity and cooperativity of oxygen binding at low pH, defined as the Root effect (Root 1931), corresponds to an exaggerated Bohr effect. At acidic pH, hemoglobin is only partially oxygenated, even at pressures of several atmospheres of pure oxygen. The Root effect was observed in intact erytrocytes, "stripped" hemolysate, Hb 1 and Hb 2 (Fig. 6), as indicated by the decrease in oxygenation at lower pH, even in the presence of the ligand at atmospheric pressures. The modulating effect of inositol hexakisphosphate,

Table 1. Sequence identity (%) in α- and β-chains of fish hemoglobins

Species	*C. clarkii*	*C. carpio*	*C. auratus*	*S. irideus* Hb IV	*S. irideus* Hb I	*T. thynnus*	*N.cor.negl.* Hb 2
(α-Chains)							
N.cor.negl. Hb 1	58	59	61	57	55	73	63
Hb 2	58	63	65	63	62	68	
T. thynnus	66	66	67	60	65		
S. irideus Hb I	65	66	67	60			
Hb IV	63	63	68				
C. auratus	77	80					
C. carpio	91						

	C. auratus	*C. carpio*	*S. irideus* Hb IV	*S. irideus* Hb I	*T. thynnus*
(β-Chains)					
N.cor.negl. Hb1, Hb2	58	57	63	53	66
T. thynnus	62	60	61	51	
S. irideus Hb I	64	64	59		
Hb IV	78	73			
C. carpio	91				

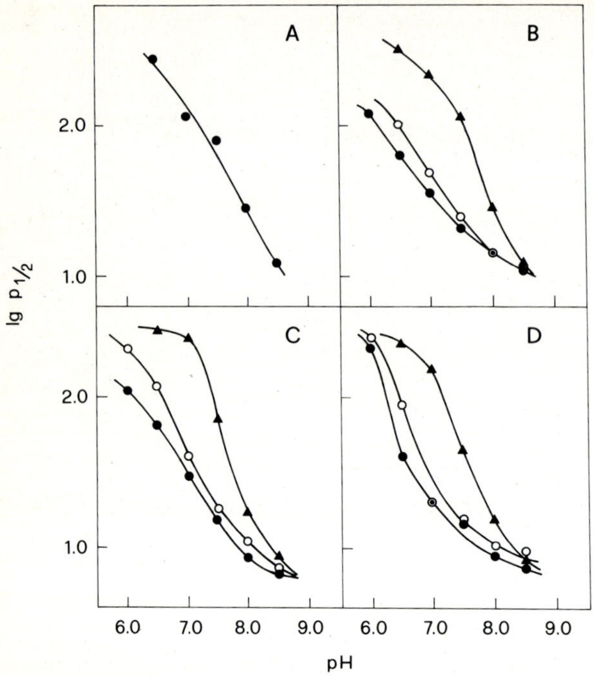

Fig. 5A-D. Oxygen equilibrium isotherms as a function of pH. (●) 0.1 M Tris or Bistris buffer, containing 0.1 M NaCl (○) and 0.1 M NaCl, 3 mM inositol hexakisphosphate (▲). $p_{1/2}$ is expressed in mmHg. Other details are given in Section 2. **A** Erythrocytes (in isotonic buffer); **B** "stripped hemolysate"; **C** Hb 1; **D** Hb 2

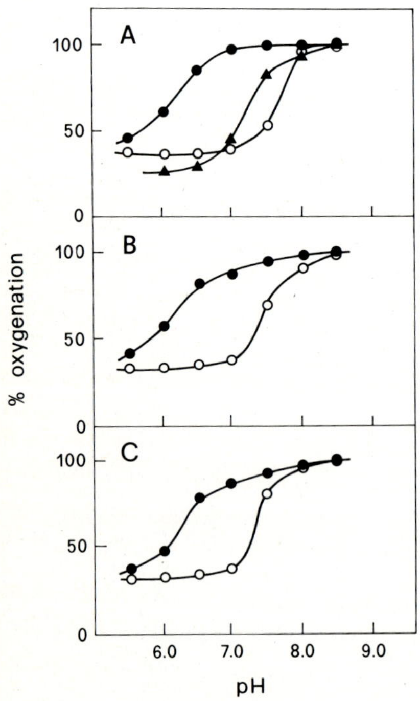

Fig. 6A-C. Oxygen saturation as a function of *pH* (Root effect). For experimental conditions, see Fig. 5 and Section 2. **A** Erythrocytes in isotonic buffer (▲); "stripped" hemolysate in the absence (●) and presence (○) of 3 mM inositol hexakisphosphate. **B** Hb 1; **C** Hb 2, in the absence (●) and presence (○) of inositol hexakisphosphate

which induced the decrease in oxygenation at higher pH values, is also evident and is similar to the effect of endogenous ATP, the presence of which accounted for the shift toward a higher pH of the inflection in the erythrocyte oxygenation curve.

4 Discussion

The first teleosts appeared in the Triassic, over 200 Ma and continued to evolve through the Jurassic and the Cretaceous. The first fossil records of most percoid families appear in the Eocene. The suborder Notothenioidei (97% endemic), which comprises six families, probably originated in Antarctic waters in the lower Tertiary. Fish of this suborder account for 109 of the 203 identified species of Antarctic fish; 52 species belong to Nototheniidae, one of the six families of Notothenioidei (Table 2). The extremely high endemism and the modest diversification in families and species is most probably due to the long isolation of the region and to the increasing harshness of climatic conditions during the cooling process south of the Convergence.

The development of cold adaptation has turned the Antarctic ocean into being now the ideal habitat for the fish fauna, by virtue of the evolutionary response of organisms at different levels of life organization (organ, cell, molecule) to the many constraints of this habitat. In this context, the peculiar hematological parameters of Antarctic fish have prompted the investigation on structure and function of hemoglobin.

The blood of *N. coriiceps neglecta*, a species commonly found in the waters of the Antarctic Peninsula, contains a major component (Hb 1) and a second one (Hb 2), present in lesser amounts. Both hemoglobins have been purified and charaterized.

The presence of a major hemoglobin component has been observed in 14 additional species of Nototheniidae; in 12 of these, Hb 2 has also been found in similar proportions (D'Avino and di Prisco 1988, 1989; di Prisco 1988). Six species of other families of Notothenioidei (Bathydraconidae, Harpagiferidae, Artedidraconidae) have a single hemoglobin. Thus, the hemoglobin multiplicity is very low: in fact, there is the clear trend of having one major component, often

Table 2. The families of the suborder Notothenioidei

Family	No. of species
Bovichthyidae	1
Nototheniidae	52
Bathydraconidae	16
Harpagiferidae	5
Artedidraconidae	19
Channichthyidae	16

accompanied by a relatively minor one, in agreement with the findings of Wells et al. (1980). In contrast, non-Antarctic fish have multiple components (Riggs 1970) that in fast swimmers, such as trout (Binotti et al. 1971), often show functional differences in oxygen binding, thus enabling oxygen transport to tissues even under conditions of extreme acidosis. In some instances, however, all components, functionally indistinguishable from one another, have pH-dependent oxygen affinities; *C. carpio*, a bottom feeder, is one example (Gillen and Riggs 1972; Tan et al. 1972).

The results on regulation of oxygen binding showed that both hemoglobins of *N. coriiceps neglecta* display a large Bohr effect. The Root effect, indicating overstabilization of the low-affinity T-state conformation of the hemoglobin molecule, was also observed. A thermodynamic analysis of CO binding had shown (Brittain 1984) that hemoglobin from the Antarctic teleost *Pagothenia borchgrevinki* closely followed the two-state model of cooperative interaction (Monod et al. 1965). Root-effect hemoglobins are usually associated with the presence of at least one of two anatomical structures which require the release of considerable amounts of oxygen: the swim bladder and the choroid rete. *N. coriiceps neglecta*, similar to all Antarctic fish, lacks a swim bladder; however, the presence of the choroid rete, associated with the eye and supplying high oxygen tension to the poorly vascularized retina (Wittenberg and Wittenberg 1974), is consistent with the availability of Root-effect hemoglobins.

Hb 1 and Hb 2 appeared to be functionally indistinguishable from each other. Preliminary results on purified components from several other species indicate this feature to be of general occurrence in Antarctic teleosts (mostly sedentary bottom feeders) and therefore related to the habitat. Multiplicity and functional differences seem to indicate a need to adapt to environmental changes; since the physicochemical features of the Antarctic marine environment are relatively constant, fish probably do not need more than one hemoglobin type. It should be noted, however, that a few species of the family Nototheniidae are not bottom feeders: *Pleuragramma antarcticum* is often found near the surface, *Dissostichus mawsoni* is a midwater predator, the cryopelagic *Pagothenia borchgrevinki* and *Trematomus newnesi* feed underneath the pack and fast ice. It would be of great interest to examine correlations in these species between the behavior patterns and the multiplicity and functional properties of hemoglobins.

The primary structure of the two hemoglobins of *N. coriiceps neglecta* has been completely elucidated: they have the β-chain in common and differ by the α-chain (D'Avino and di Prisco 1989).

N-Acetylated Ser was detected at the α-chain N-terminus of both Hb 1 and Hb 2; Val was the N-terminal residue of the β-chain in common to the two hemoglobins. The residues in the five positions (four are heme contacts) known to be invariant in the globin chains of all vertebrates (Dickerson and Geis 1983) are conserved. Nine of the other 15 α-chain heme contacts are also conserved. Differences observed in Antarctic and non-Antarctic teleosts in five (E11, FG3, FG5, H12, H15) of the other six positions are all conservative; in F7, the only heme contact in which the α-chain of *N. coriiceps neglecta* Hb 1 differs from all others (including Hb 2 of the same species), hydrophilic Gln replaces hydrophobic Leu. In β-chains, 13 of the 18 heme contacts are conserved; in the first of the other five positions (C7, CD3, CD4, F7, H19), positively charged His (also found in trout Hb IV) replaces a hydrophobic residue (Tyr, Phe).

Among the 23 positions (partially coinciding with the heme contacts) previously considered to be invariant in all α-chains, two are changed in *N. coriiceps neglecta* Hb 1 and Hb 2: in B8 and E14, Glu and Ala are replaced by Asp and Gly, respectively. In addition, the replacement in Hb 1 of Asp by Asn in H9, located in one of the $\alpha_1\beta_1$ (or $\alpha_2\beta_2$) "packing contact" regions, may well be responsible for structural changes. The 18 positions considered invariant in all β-chains were all conserved in *N. coriiceps neglecta*.

In considering the 26 identical residues in sperm whale myoglobin and in α- and β-chains of horse and human hemoglobin (Dickerson and Geis 1983), each one having an important function in the secondary and tertiary structure, six (NA1, A15, CD4, CD7, G16, HC1) and five (NA1, CD4, E5, FG2, G16) substitutions were found in the α-chain of Hb 1 and Hb 2, respectively; with the exception of FG2 (Thr replaces Lys), the replacements are not likely to alter the structure significantly. The β-chain of the Antarctic teleost features eight replacements (NA3, A8, A12, A14, A15, E5, G16, HC1); in E5 and HC1, where Ala and Gln replace Lys, a structural alteration may have occurred.

In both chains, a very high degree of conservation is evident in the sequence positions involved in the $\alpha_1\beta_2$ (or $\alpha_2\beta_1$) "sliding contact" regions (where movement occurs when the heme binds a ligand such as oxygen), mainly located in helices C and G and in the FG corner (C2, C3, C5, C6, C7, CD2, FG3, FG4, FG5, G1, G2, G3, G4, G7, HC2, HC3): mutations in these domains would clearly have a disturbing effect on the functional features. The replacement in CD2 of both α-chains (Ser in place of Pro, similar to trout, goldfish, and tuna) might alter the interactions involving His at FG4 in the deoxy configuration.

On the other hand, in the "packing contacts" of both chains (where the change from the deoxy to the oxy configuration does not alter the $\alpha_1\beta_1$ or $\alpha_2\beta_2$ packing), mainly located in helices B, G, and H and in the GH corner (B12, B15, B16, C1, G6, G10, G14, G17, G18, GH2, GH3, GH5, H1, H2, H3, H5, H6, H9, H10), mutations have occurred more frequently. In fact, dissimilarities have been found among each of the three new sequences and the others in more than half of the 19 residues; some of these substitutions are nonconservative.

The examination of the sequences of Figs. 2 and 4 reveals further differences in both chains. Table 3 summarizes replacements introducing changes in charge and/or in (possibly) structure between *N. coriiceps neglecta* and at least four non-Antarctic species.

The primary structure of this protein, and in particular the domains of structural importance, have been highly conserved during evolution, as indicated also by the degree of sequence identities, summarized in Table 1. In fact, most of the replacements in the sequences appear to be localized in domains where the structural requirements for the biological function are less stringent.

The sequence of *N. coriiceps neglecta* Hb 1 and Hb 2 shows the highest similarity with bluefin tuna hemoglobin. *T. thynnus* belongs to the same order, Perciformes, or, according to some sources, to a closely related one, Thun-

Table 3. Replacements of some residues other than the invariant ones, the heme contacts, or those involved in "sliding contact" and "packing contact" regions

Chain	Sequence differences		Position
	N.cor. negl.	Non-Antarctic sp.	
α (Hb 1)	Ala	Asp,Glu	B4
	Ser	Asp,Glu	CD6
	-His-Pro-Asp-Ile-	-Ser-Gly-Pro-Val-	E1–4
	Asn	Asp	EF3,H9
(Hb 2)	Thr	Asp	A2
	Glu	Ala,Ser	A6
	Lys	Ala,Pro,Ser	CD5
	-Met-Thr-	-Lys-Lys-	E9–10
	Lys	Gly	E13
	Ala	Pro	H2
	Arg	Leu,Ala	E17
(Both)	Asn	Ala	B7
	Ser	Gly	B11
	Thr	Ser	CD8
β	Asn	Asp,Glu	NA2
	Ala	Ser	A6
	Asp	Ala,Gly	A10
	Asp	Asn	B1
	Ser	Ala	B11
	Tyr	Ser	CD8
	Ala	Pro	D2,E2
	Asn	Lys	E3
	Lys	Asn	EF4
	Asp	Ala,Asn	EF7

niformes, and is therefore phylogenetically less remote from the nototheniid than the others. The sequence of the α-chain of Hb 1 is more similar to that of the α-chain of bluefin tuna hemoglobin than to the α-chain of Hb 2. This peculiarity exists, to a larger extent, in trout α- and β-chain sequences (Petruzzelli et al. 1984, 1989): trout Hb IV is more similar to carp hemoglobin than to trout Hb I. However, trout Hb IV, unlike Hb I, is also functionally similar to carp hemoglobin, whereas *N. coriiceps neglecta* Hb 2 is functionally indistinguishable from the major component Hb 1 and its physiological role remains unclear. It is conceivable that the relatively minor component Hb 2 may simply be an evolutionary remnant.

X-ray crystallography and model building have suggested that several residues, especially of the β-chain, have a key role in the functional properties, in relation to regulation of oxygen affinity by pH and allosteric effectors, alterations of subunit cooperativity, transitions from R- to T-state and *vice versa* (Arnone 1972; Barra et al. 1981; Perutz and Brunori 1982). Root and Bohr effects appear to be linked to the presence of several polar residues in the β-chain sequence: Lys EF6, Ser F9, Glu FG1, Arg H21, His HC3. With the exception of the first residue, replaced by Val, the others are present in the chains of *N. coriiceps neglecta* Hb

1 and Hb 2 (in H21 Arg is conservatively replaced by Lys). In the deoxy T-structure, Ser F9 donates a hydrogen bond to the unbound oxygen atom of His HC3 and accepts a hydrogen bond from the peptide -NH- of the same residue, thereby stabilizing the C-terminal salt bridges (respectively, imidazole and -COOH of His HC3 with Glu FG1 and, in the α-chain, Lys C5) in the T-structure. Thus, His HC3 is no longer available for the salt bridge formed with Lys HC1 in mammals in the R-state. As an overall result, in Root-effect hemoglobins, the T-state is stabilized and the R-state destabilized. In trout Hb I, which has neither effect, all five polar residues (present in trout Hb IV) have been replaced by apolar ones, and Lys HC1 has been replaced by Arg, still able to form the salt bridge with the C-terminus and stabilize the R-quaternary conformation. On the other hand, in Root-effect trout Hb IV and *N. coriiceps neglecta* hemoglobins, Gln HC1 cannot form the latter salt bridge and provide such stabilization and the strong interaction between His HC3 and Glu FG1 in the transition from the R- to the T-state (Perutz and Brunori 1982) is energetically favored (in trout Hb I, in FG1) Glu is replaced by Asn).

Instead of D-2,3-diphosphoglycerate (one of the ligands involved in the heterotropic interactions in mammal hemoglobins), ATP and, to a lesser extent, GTP are among the allosteric effectors used by teleosts to regulate oxygen binding. Perutz and Brunori (1982) suggested differences in the stereochemistry of the respective organophosphate binding sites, related to the replacements of one or more of three residues coating the central crevice between the two β-chains (Arnone 1972) of the diphosphoglycerate site in mammals: His NA2, Lys EF6 and His H21. In trout Hb IV, Lys EF6 is conserved, and the other two residues are Asp NA2 and Arg H21: this stucture may explain the preferential binding of ATP and GTP. (None of the 3 positive charges is conserved in trout Hb I, which has Glu NA2, Leu EF6 and Ser H21 and does not respond to the regulators.) In *N. coriiceps neglecta* Hb 1 and Hb 2, the oxygen affinity is lowered by ATP and the residues are Asn NA2, Val EF6, and Lys H21, suggesting firstly that binding of heterotropic ligands does require a positively charged residue at H21 and, secondly, that the presence of one positive and one negative charge in the other two positions (trout Hb IV) does not exclude an alternative stereochemical configuration of the A(G)TP site, possibly mediated by some of the other replacements in the primary structure of one or both chains.

Acknowledgements. This work is in the framework of the Italian National Programme for Antarctic Research. It was partially supported by grant DPP 82-18356 from the Division of Polar Programs, National Science Foundation, Washington, D.C., USA. Fast-atom-bombardment mass spectrometry was performed at the Mass Spectrometry Center, Faculty of Medicine, University of Florence, Italy.

References

Adie RJ (1970) Past environment and climates of Antarctica. In: Holdgate MW (ed) Antarctic ecology, vol I. Academic Press, London, pp 7–14

Arnone A (1972) X-Ray diffraction study of binding of 2,3-diphosphoglycerate to human deoxyhaemoglobin. Nature 237:146–149

Barker PF, Burrell J (1977) The opening of the Drake Passage. Mar Geol 25:15–34

Barra D, Bossa F, Brunori M (1981) Structure of binding sites for heterotropic effectors in fish hemoglobins. Nature 293:587–588

Barra D, Petruzzelli R, Bossa F, Brunori M (1983) Primary structure of hemoglobin from trout (*Salmo irideus*). Amino acid sequence of the β-chain of trout Hb I. Biochim Biophys Acta 742:72–77

Binotti I, Giovenco S, Giardina B, Antonini E, Brunori M, Wyman J (1971) Studies on the functional properties of fish hemoglobins. II. The oxygen equilibrium of the isolated hemoglobin components from trout blood. Arch Biochem Biophys 142:274–280

Bossa F, Barra D, Petruzzelli R, Martini F, Brunori M (1978) Primary structure of hemoglobin from trout (*Salmo irideus*). Amino acid sequence of α-chain of Hb trout I. Biochim Biophys Acta 536:298–305

Brauer AW, Oman CL, Margolies MN (1984) Use of *o*-phthalaldehyde to reduce background during automated Edman degradation. Anal Biochem 137:134–142

Braunitzer G, Rodewald K (1980) Die Sequenz der α- und β-ketten des Hämoglobins des Goldfisches (*Carassius auratus*). Hoppe-Seyler's Z Physiol Chem 361:587–590

Brittain T (1984) A two-state thermodynamic and kinetic analysis of the allosteric functioning of the haemoglobin of an extreme poikilotherm. Biochem J 221:561–568

Coburn CB, Fischer BA (1973) Red blood cell hematology of fishes: a critique of techniques and a compilation of published data. J Mar Sci 2:37–58

Craddock C (1970) Antarctic geology and Gondwanaland. Antarct J US 5(3):53–57

D'Avino R, di Prisco G (1988) Antarctic fish hemoglobin: an outline of the molecular structure and oxygen binding properties. I. Molecular structure. Comp Biochem Physiol 90B:579–584

D'Avino R, di Prisco G (1989) Hemoglobin from the Antarctic fish *Notothenia coriiceps neglecta*. 1. Purification and characterization. Eur J Biochem 179:699–705

D'Avino R, Caruso C, Romano M, Camardella L, Rutigliano B, di Prisco G (1989) Hemoglobin from the Antarctic fish *Notothenia coriiceps neglecta*. 2. Amino acid sequence of the α-chain of Hb 1. Eur J Biochem 179:707–713

Denton GH, Armstrong RL, Stuiver M (1970) Late Cenozoic glaciation in Antarctica. Antarct J US 5(1):15–21

Dickerson RE, Geis I (1983) Hemoglobin: structure, function, evolution and pathology. Benjamin/Cummings, Menlo Park, CA

Dietz RS, Sproll WP (1970) Fit between Africa and Antarctica: a continental drift reconstruction. Science 167:1612–1614

di Prisco G (1988) A study of hemoglobin in Antarctic fishes: purification and characterization of hemoglobins from four species. Comp Biochem Physiol 90B:631–635

di Prisco G, D'Avino R (1989) Molecular adaptation of the blood of Antarctic teleosts to environmental conditions. Antarct Sci 1:119–124

di Prisco G, Giardina B, D'Avino R, Condo' SG, Bellelli A, Brunori M (1988) Antarctic fish hemoglobin: an outline of the molecular structure and oxygen binding properties. II. Oxygen binding properties. Comp Biochem Physiol 90B:585–591

Drapeau GR (1977) Cleavage at glutamic acid with staphylococcal protease. Methods Enzymol 47:189–194

Everson I, Ralph R (1968) Blood analyses of some Antarctic fish. Br Antarct Surv Bull 15:59–62

Frakes LA, Crowell JC (1970) Geologic evidence for the place of Antarctica in Gondwanaland. Antarct J US 5(3):67–69

Giardina B, Amiconi G (1981) Measurement of binding of gaseous and nongaseous ligands to hemoglobins by conventional spectrophotometric procedures. Methods Enzymol 76:417–427

Gillen RG, Riggs A (1972) Structure and function of the hemoglobins from the carp, *Cyprinus carpio*. J Biol Chem 247:6039–6046

Gross E, Witkop B (1961) Selective cleavage of the methionyl peptide bonds in ribonuclease with cyanogen bromide. J Am Chem Soc 83:1510

Grujić-Injac B, Braunitzer G, Stangl A (1980) Hemoglobins, XXXV: the sequence of the β_A- and β_B-chains of the hemoglobins of the carp (*Cyprinus carpio*). Hoppe-Seyler's Z Physiol Chem 361:1629–1639

Hilse K, Braunitzer G (1968) Die Aminosäuresequenz der α-Ketten der beiden Hauptkomponenten des Karpfenhämoglobins. Hoppe-Seyler's Z Physiol Chem 349:433–450

Hureau J-C, Petit D, Fine JM, Marneux M (1977) New cytological, biochemical and physiological data on the colorless blood of the Channichthyidae (Pisces, Teleosteans, Perciformes). In: Llano GA (ed) Adaptations within Antarctic ecosystems. Smithsonian Institution, Washington, pp 459–477

Kennett JP (1977) Cenozoic evolution of Antarctic glaciation, the circum-Antarctic ocean and their impact on global paleoceanography. J Geophys Res 82:3843–3876

King LC (1958) Basic palaeogeography of Gondwanaland during the late Paleozoic and Mesozoic eras. Q J Geol Soc Lond 114:47–77

Larsson A, Johansson-Siöbeck ML, Fänge R (1976) Comparative studies of some haematological and biochemical blood parameters in fishes from the Skagerrak. J Fish Biol 9:425–440

Le Masurier WE (1970) Volcanic evidence for early Tertiary glaciation in Marie Byrd Land. Antarct J US 5(5):154–155

Monod J, Wyman J, Changeux JP (1865) On the nature of allosteric transitions: a plausible model. J Mol Biol 12:88–118

Perutz MF (1969) The haemoglobin molecule. Proc R Soc Lond, Ser B 173:113–140

Perutz MF, Brunori M (1982) Stereochemistry of cooperative effects in fish and amphibian haemoglobins. Nature 299:421–426

Petruzzelli R, Barra D, Goffredo BM, Bossa F, Coletta M, Brunori M (1984) Amino acid sequence of β-chains of hemoglobin IV from trout (*Salmo irideus*). Biochim Biophys Acta 789:69–73

Petruzzelli R, Barra D, Sensi L, Bossa F, Brunori M (1989) Amino acid sequence of α-chain of hemoglobin IV from trout (*Salmo irideus*). Biochim Biophys Acta 995:255–258

Powers DA, Edmundson AB (1972) Multiple hemoglobins of catostomid fish. II. The amino acid sequence of the major α-chain from *Catostomus clarkii* hemoglobins. J Biol Chem 247:6694–6707

Riggs A (1970) Properties of fish hemoglobins. In: Hoar WS, Randall DJ (eds) Fish physiology, vol IV. Academic Press, New York, pp 209–252

Rodewald K, Obertür W, Braunitzer G (1987) Homeothermic fish and hemoglobin: primary structure of the hemoglobin from bluefin tuna (*Thunnus thynnus*, Scombroidei). Biol Chem Hoppe-Seyler 368:795–805

Root RW (1931) The respiratory function of blood in marine organisms. Biol Bull Mar Biol Lab, Woods Hole 61:427–456

Ruud JT (1954) Vertebrates without erythrocytes and blood pigment. Nature 173:848–850

Schinina' ME, De Biase D, Bossa F, Barra D (1988) In situ treatment of proteins prior to sequence analysis. J Protein Chem 7:284–286

Schopf JM (1970) Gondwana paleobotany. Antarct J US 5(3):62–66

Somero GN, DeVries AL (1967) Temperature tolerance of some Antarctic fishes. Science 156:257–258

Tan AL, De Young A, Noble RW (1972) The pH dependence of the affinity, kinetics, and cooperativity of ligand binding to carp hemoglobin, *Cyprinus carpio*. J Biol Chem 247:2493–2498

Tanner WF (1968) Tertiary sea level symposium – Introduction. Paleogeogr Paleoclimatol Paleoecol 5:7–14

Tetens V, Wells RMG, DeVries AL (1984) Antarctic fish blood: respiratory properties and the effects of thermal acclimation. J Exp Biol 109:265–279

Wells RMG, Jokumsen A (1982) Oxygen binding properties of hemoglobins from Antarctic fishes. Comp Biochem Physiol 71B:469–473

Wells RMG, Ashby MD, Duncan SJ, Macdonald JA (1980) Comparative studies of the erythrocytes and haemoglobins in nototheniid fishes from Antarctica. J Fish Biol 17:517–527

Wilson JT (1963) Continental drift. Sci Am 208:86–100

Wittenberg JB, Wittenberg DK (1974) The choroid rete mirabile. I. Oxygen secretion and structure: comparison with the swimbladder rete mirabile. Biol Bull 145:116–136

Wohlschlag DE (1964) Respiration metabolism and ecological characteristics of some fishes in McMurdo Sound, Antarctica. Antarct Res Ser 1:33–62

Cold-Stable Microtubules from Antarctic Fish

H. W. Detrich III

1 Introduction

The cytoplasmic microtubules of eukaryotic cells participate in many fundamental processes, including mitosis, nerve growth and regeneration, the determination of cell shape, and the transport of organelles within cells (Dustin 1984). The assembly of microtubules from their major subunit proteins, tubulin $\alpha\beta$-dimers and microtubule-associated proteins (MAPs), is an entropically driven reaction favored by high temperatures and mediated by the release of structured water from sites of interdimer contact (Correia and Williams 1983). Thus, the microtubule proteins of homeotherms (e.g., mammals and birds) form microtubules at temperatures near 37 °C, and these "cold-labile" polymers disassemble at low temperatures (0–4 °C). By contrast, the microtubules of cold-living poikilotherms, such as those found in the Antarctic marine ecosystem, must assemble and function at temperatures as low as –1.9 °C. However, relatively little is known regarding the polymerization of microtubule proteins from organisms adapted to low body temperatures.

To learn about the molecular adaptations that are responsible for microtubule assembly in poikilotherms, my laboratory has chosen the coastal fish of the Antarctic Peninsula as its model system. These cold-adapted fish diverged from temperate osteichthyans approximately 40 million years ago, concomitant with the isolation of the Antarctic continent in a cooling southern ocean (DeWitt 1971). Near Palmer Station, Antarctica, these stenothermal fish live at an annual mean bottom temperature of –1.0 °C (seasonal range –0.1 °C in January to –1.9 °C in August; Lowry 1975). Thus, Antarctic fish constitute a unique fauna that is well suited to the study of cold-stable microtubules.

The formation of microtubules at low temperatures may result from altered tubulins, from unique microtubule-associated, cold-stabilizing factors, or from both. We (Detrich et al. 1989) and others (Williams et al. 1985) have shown previously that pure, MAP-free brain tubulins from Antarctic fish polymerize efficiently at the low body temperatures (–1.8 to +2 °C) experienced by these extreme poikilotherms. Therefore, the major assembly-enhancing adaptations of the Antarctic fish microtubule proteins reside in their tubulins, not in their MAPs. Here, I review studies from my laboratory of the polymerization energetics of

Department of Biology, Northeastern University, 360 Huntington Avenue, Boston MA 02115, USA.

Guido di Prisco (Ed)
Life Under Extreme Conditions

Antarctic fish tubulins (Detrich et al. 1989) and of the structures of the component α- and β-tubulin subunits (Detrich et al. 1987). The results suggest that the functional adaptations of these tubulins include increases in the proportion or in the strength (or both) of the hydrophobic interactions that mediate microtubule assembly, and that many of the adaptations are present in their unique α-chains.

2 Purification of Brain Tubulins from Antarctic Fish

We have purified tubulins from the brain tissues of three species of Antarctic fish (two nototheniids, *Notothenia gibberifrons* and *N. coriiceps neglecta*, and one channichthyid, *Chaenocephalus aceratus*) by diethylaminoethyl (DEAE) ion-exchange chromatography and one cycle of microtubule assembly in vitro, as described elsewhere (Detrich and Overton 1986; Detrich et al. 1989). Figure 1 shows a sodium dodecyl sulfate (SDS) urea-polyacrylamide gradient gel that contains protein fractions that were obtained during preparation of tubulin from *N. gibberifrons*. The microtubules (lane 7) obtained by this method are composed of the α- and β-tubulins and are free of MAPs. Based on the sensitivity of the Coomassie stain, we estimate that these preparations contain greater than 98% α- and β-tubulins. Furthermore, the yield of tubulin (1.2 mg/g brain tissue) achieved by this protocol is approximately tenfold larger than that produced by a different method involving two cycles of temperature-dependent microtubule assembly/disassembly and phosphocellulose chromatography (Williams et al. 1985).

3 Polymerization of Antarctic Fish Tubulins

3.1 Assembly of Microtubules

To evaluate the functional properties of the Antarctic fish tubulins, we (Detrich et al. 1989; Himes and Detrich 1989) have employed turbidimetry (cf. Gaskin et al. 1974), electron microscopy, and a quantitative sedimentation assay (cf. Johnson and Borisy 1975). Important results are presented in Figs. 2–5.

Figure 2 shows the polymerization of *N. gibberifrons* tubulin following temperature jumps from 0 °C to final temperatures of 5, 10, or 20 °C. Following a short lag, the turbidity of the sample (0.6 mg/ml) at 20 °C increased and reached a plateau value approximately 10 min after the start of assembly. Qualitatively similar polymerization profiles were observed at 5 °C (tubulin concentration = 1.3 mg/ml) and at 10 °C (tubulin concentration = 0.96 mg/ml; Fig. 2), but these reactions proceeded more slowly and were accompanied by small overshoots in turbidity (cf. Detrich et al. 1985). Cooling of polymerized samples to 0 °C caused their solution turbidities to decrease to near the zero-time values

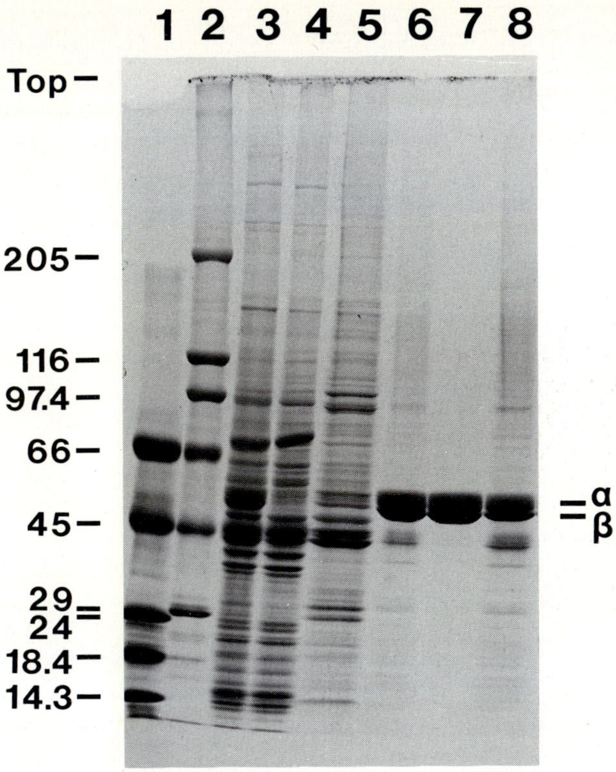

Fig. 1. Purification of brain tubulin from the Antarctic fish, *Notothenia gibberifrons*. Tubulin was purified from brain tissue by DEAE ion-exchange chromatography and one round of microtubule polymerization as described previously (Detrich and Overton 1986; Detrich et al. 1989). Protein fractions obtained at various stages of the purification were examined by electrophoresis on an SDS urea-polyacrylamide gradient gel. *Lanes: 1* low-molecular-weight standards, including bovine serum albumin [relative molecular mass (M_r) = 66 000], ovalbumin (M_r = 45 000), trypsinogen (M_r = 24 000), β-lactoglobulin (M_r = 18 400), and lysozyme (M_r = 14 300); *2* high-molecular-weight standards, including myosin (M_r = 205 000), β-galactosidase (M_r = 116 000), phosphorylase *b* (M_r = 97 400), bovine serum albumin, ovalbumin, and carbonic anhydrase (M_r = 29 000); *3* high-speed supernatant from *N. gibberifrons* brain; *4* proteins present in the flow-through fraction following application of the high-speed supernatant to a column of DEAE-Sephacel; *5* and *6* proteins released from the ion-exchange resin by PMTG buffer [0.1 M piperazinediethanesulfonic acid-NaOH (Pipes-NaOH), 1 mM $MgSO_4$, 1 mM *p*-tosyl-L-arginine methyl ester HCl, 0.1 mM guanosine 5′-triphosphate (GTP), pH 6.9 at 20 °C] containing 0.15 and 0.40 M NaCl, respectively; *7* and *8* pellet and supernatant obtained by polymerization of sample 6 (20 °C, 20 min) followed by centrifugation (40 000 × g, 20 °C, 20 min) to collect the microtubule polymer. Electrophoretic migration was from top to bottom. The molecular weights of the standards (in thousands) and the positions of the tubulin chains (α and β) and of the top of the gel are indicated on the vertical axes. (Reprinted with permission from Detrich et al. 1989. Copyright 1989 American Chemical Society)

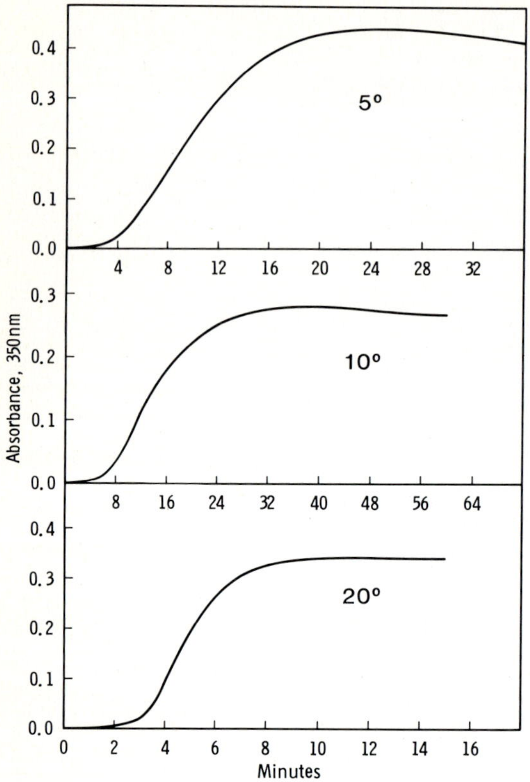

Fig. 2. Polymerization of tubulin from an Antarctic fish at three temperatures. Samples of tubulin from *N. gibberifrons* [in PME buffer (0.1 M Pipes-NaOH, 1 mM $MgSO_4$, 1 mM ethylene glycol bis(β-aminoethyl ether)-*N,N,N′,N′*-tetraacetic acid, pH 6.9 at 20 °C) containing 0.1 mM GTP, 10 mM acetyl phosphate, and 0.28 units/ml acetate kinase] were warmed from 0 °C to final temperatures of 5, 10, or 20 °C at zero time, and microtubule assembly was monitored by turbidimetry. The development of turbidity (apparent absorbance at 350 nm) is plotted as a function of time after warming. Tubulin concentrations: 5 °C, 1.3 mg/ml; 10 °C, 0.96 mg/ml; 20 °C, 0.60 mg/ml. (Reprinted with permission from Himes and Detrich 1989. Copyright 1989 American Chemical Society)

(Detrich et al. 1989). Together, these results are consistent with the reversible formation of microtubule polymer in vitro in response to changes in temperature.

Figure 3 shows a representative negative-stain electron micrograph of the polymers formed when a solution of tubulin from *N. coriiceps neglecta* was warmed from 0 to 20 °C. The predominant products of assembly at this non-physiological temperature were microtubules of normal morphology and diameter; the protofilaments of these microtubules are readily apparent (Fig. 3). Some microtubule polymorphs (e.g., protofilament sheets and "hooked" microtubules), which may be intermediates on the assembly pathway to microtubules, were also observed (for micrographs, see Detrich et al. 1989). When polymerization was initiated at the physiological temperature of 0 °C by addition

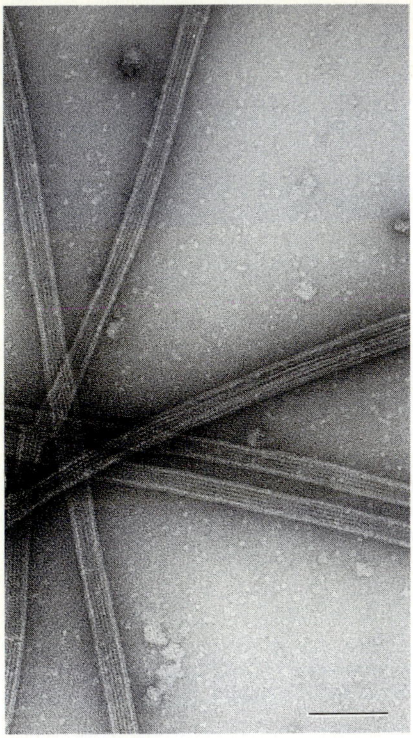

Fig. 3. Electron micrograph of microtubule polymer assembled in vitro from an Antarctic fish tubulin. A solution of *N. coriiceps neglecta* tubulin [0.64 mg/ml in PME buffer (see legend to Fig. 2) containing 1 mM GTP] was warmed from 0 to 20 °C at zero time, and a negatively stained specimen was prepared 30 min after the start of assembly. *Bar* = 100 nm. (Reprinted with permission from Detrich et al. 1989. Copyright 1989 American Chemical Society)

of guanosine 5′-triphosphate (GTP) to a solution of *N. gibberifrons* tubulin (4.4 mg/ml) lacking the nucleotide, microtubules again constituted the majority of the polymer. We conclude that MAP-free brain tubulins from Antarctic fish, like those from homeothermic vertebrates (mammals and birds; Lee and Timasheff 1975, 1977), possess the structural information necessary to direct microtubule assembly in vitro.

3.2 Temperature Dependence of Microtubule Assembly

Microtubule assembly systems in vitro are characterized by a minimal, or "critical", concentration of tubulin below which polymerization does not occur (Gaskin et al. 1974; Johnson and Borisy 1975). Above the critical concentration, the extent of assembly is a linear function of the total tubulin concentration. We measured the critical concentrations for polymerization of tubulins from two Antarctic fish species at temperatures between 0 and 18 °C by a modification of the quantitative sedimentation assay of Johnson and Borisy (1975; for details, see Detrich et al. 1989). As shown in Fig. 4, values for the critical concentrations decreased from 0.87 mg/ml at 0 °C to 0.02 mg/ml at 18 °C. For comparison, critical concentration values for pure mammalian brain tubulins at the same

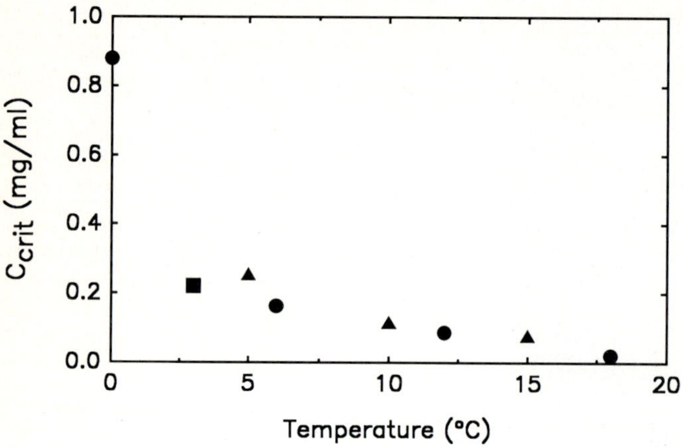

Fig. 4. Temperature dependence of the critical concentration for polymerization of Antarctic fish tubulins. Critical concentrations (C_{crit}) were measured at temperatures between 0 and 18 °C by means of a quantitative sedimentation assay (Johnson and Borisy 1975; Detrich et al. 1989). Three tubulin preparations from two Antarctic fish species were used in these studies: *circles*, *N. gibberifrons* tubulin, preparation 1; *triangles*, *N. gibberifrons* tubulin, preparation 2; *square*, *N. coriiceps neglecta* tubulin

temperatures must be approximately two orders of magnitude larger (Williams et al. 1985). Our results indicate that Antarctic fish tubulins are able to form microtubules at physiological temperatures and at low protein concentrations.

Figure 5 presents the critical concentration data in the form of a van't Hoff plot. For this analysis, we have assumed that the apparent association constant for microtubule elongation (i.e., the addition of a tubulin dimer to the end of a microtubule) is equal to the reciprocal of the critical concentration (Gaskin et al. 1974; Lee and Timasheff 1977). The data points are adequately fit by a straight line ($r = -0.956$). From these data, we estimate that the standard enthalpy change, $\Delta H°$, for microtubule elongation is +26.9 kcal/mol, and the corresponding standard entropy change, $\Delta S°$, is +123 eu. Thus, the assembly of microtubules from the tubulins of Antarctic fish is strongly entropy driven.

3.3 Salt Dependence of Microtubule Assembly

Hydrophobic interactions and ionic (electrostatic) bonds make contributions of positive sign to the entropy of subunit association, whereas hydrogen bonds and van der Waals contacts make negative contributions (Ross and Subramanian 1981). Thus, the entropy-driven polymerization of Antarctic fish tubulins must be dominated by hydrophobic interactions and/or ionic bonds. To evaluate the relative contributions of the two interactions, we measured the critical concentration for polymerization of *N. gibberifrons* tubulin at 10 °C as a function of the concentration of an antichaotropic (i.e., structure-making) salt, NaCl (Detrich et al. 1989). (Such salts strengthen hydrophobic interactions and weaken ionic

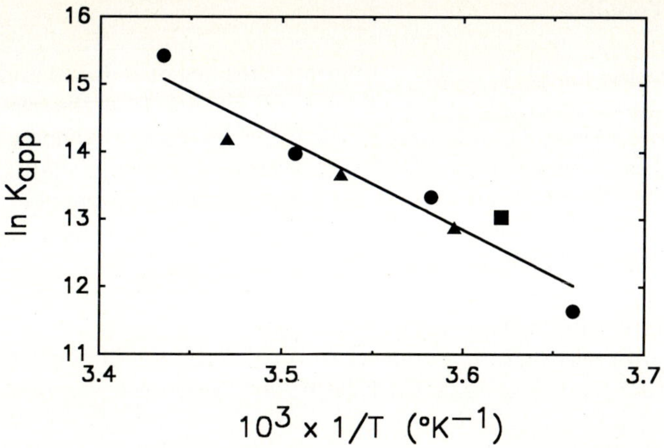

Fig. 5. Thermodynamics of the polymerization of Antarctic fish tubulins in vitro: van't Hoff analysis of the apparent equilibrium constant for microtubule elongation. The natural logarithm of the apparent equilibrium constant for microtubule elongation (i.e., the natural logarithm of the reciprocal of the critical concentration) is shown as a function of the reciprocal of the absolute temperature. The best-fitting straight line through the data points was determined by linear-regression analysis. For symbols, see legend to Fig. 4. (Reprinted with permission from Detrich et al. 1989. Copyright 1989 American Chemical Society)

bonds.) The critical concentration increased moderately, from 0.041 to 0.34 mg/ml, as the NaCl concentration was raised from 0 to 0.4 M. By contrast, NaCl at concentrations greater than 0.25 M inhibits completely the polymerization of tubulins (Himes et al. 1977) or total microtubule proteins (Olmsted and Borisy 1975) from mammals. It appears, then, that hydrophobic interactions play a greater role in the polymerization of Antarctic fish tubulins than they do in the assembly of mammalian tubulins.

4 Structure of Antarctic Fish Tubulins

Our ultimate objective is to determine the structural adaptations (primary sequence changes and/or unique posttranslational modifications) that underlie the functional properties of the Antarctic fish tubulins. To address this problem, we have used a comparative strategy to identify biochemical features that are unique to the tubulins of these fish. My laboratory has shown previously that brain tubulins from these cold-adapted fish differ from those of the temperate catfish and the cow in isotubulin composition, in amino acid composition, and in net charge (Detrich and Overton 1986, 1988). Below, I describe our recent comparative studies (Detrich et al. 1987) of the heterogeneity and structures of the α- and β-tubulin subunits from Antarctic fish, from temperate fish, and from a mammal (Detrich et al. 1987).

4.1 Heterogeneity of Tubulins

Figure 6 shows an SDS polyacrylamide gel containing samples of reduced and carboxymethylated tubulins from two species of Antarctic fish and from the cow. Tubulins from *C. aceratus* (lane 1) and from *N. coriiceps neglecta* (not shown) produced two α-chains each (designated α_1 and α_2 in order of decreasing relative mobility), whereas *N. gibberifrons* tubulin (lane 2) gave three α-components (termed α_1, α_2, and α_3 as before). Bovine α-tubulin migrated as a major species with a second, poorly resolved component trailing the main band. Clearly, the major, rapidly migrating α_1-tubulin ($>$ 55% of total α) of the Antarctic fish is absent in the mammalian tubulin (lanes 3–4).

To evaluate the significance of the α_1-tubulin variant of the cold-adapted fish, we also examined tubulins from two temperate fish (the channel catfish, *Ictalurus punctatus*, and the smooth dogfish shark, *Mustelus canis*). The α-tubulins of the temperate fish were generally similar to bovine α-tubulin. Both the catfish and the dogfish possess a major α-variant that comigrated with the prominent α-chain of the cow (not shown). Catfish tubulin contains, in addition, a small amount (17%) of an "α_1-like" tubulin, whereas dogfish tubulin lacks this minor component (not shown). Thus, a major feature that distinguishes the tubulins of the cold-adapted fish from those of the temperate fish and the mammal is the quantity of their most rapidly migrating α-chains.

In contrast to the α-tubulins, the β-tubulins from the Antarctic fish, the temperate fish, and the cow were qualitatively similar. Each of the fish tubulins, whether cold-adapted or temperate, contains major and minor β-chains that comigrated electrophoretically with the β_1 (major) and β_2 (minor) tubulins of the mammal (cf. lanes 1–2 with lanes 3–4; see also Detrich et al. 1987). Quantitatively,

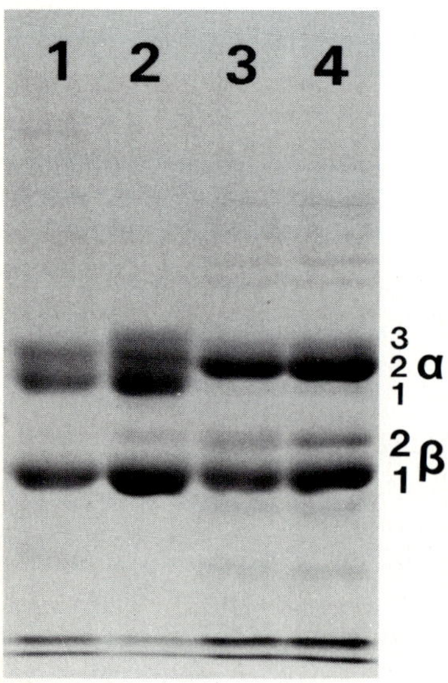

Fig. 6. Electrophoretic analysis of tubulins from two Antarctic fish and from a mammal. Following reduction and carboxymethylation (Cresfield et al. 1963), purified brain tubulins from *C. aceratus* (*lane 1*) and *N. gibberifrons* (*lane 2*) and once-cycled microtubule protein from the cow (two loadings, *lanes 3–4*) were electrophoresed in the presence of SDS on a 5.5% polyacrylamide slab gel (Detrich et al. 1987). Electrophoretic migration was from top to bottom. Only the tubulin-containing region of the gel is shown. The approximate positions of the α-tubulin variants (α_1–α_3) and of the major (β_1) and minor (β_2) β-tubulins are indicated on the vertical axis

the "β_2-like" tubulins of the fish constitute a smaller proportion (4–17%) of the total β-tubulin than does bovine β_2 (25%). The β_1/β_2 ratio does not, however, differentiate between the temperate and cold-living fish.

4.2 Tertiary Structure of β-Tubulins

The similarity of the β-tubulins extends to their native conformations in the tubulin dimer. When tubulins from two Antarctic fish (*N. gibberifrons*, *C. aceratus*), from the temperate dogfish, and from the cow were reacted with the sulfhydryl-directed cross-linking reagent, *N,N'*-ethylenebis(iodoacetamide), each produced three cross-linked derivatives of β_1-tubulin, designated β_1^s, β_1^*, and β_1^{s*} according to the nomenclature of Roach and Ludueña (1984). Thus, we conclude that β_1-tubulins from these organisms possess reactive cysteines located in regions of comparable three-dimensional structure.

4.3 Structural Analysis of α- and β-Tubulins by Peptide Mapping

The results of our electrophoretic analyses, together with other data (see Detrich and Overton 1986, 1988; Detrich et al. 1987), suggest that the α-tubulins of the Antarctic fish differ structurally from those of temperate fish and a mammal, whereas the β-tubulins from these organisms may be quite similar. To evaluate this hypothesis, we compared the α- and β-tubulins of the three Antarctic fish, of the temperate dogfish, and of the cow by one-dimensional peptide mapping. Preparations of α- and β-subunits minimally contaminated with the complementary chains were isolated by preparative electrophoresis of reduced and carboxymethylated tubulins on SDS-polyacrylamide gels. Figure 7 presents peptide maps of the five α- and five β-tubulins after chemical cleavage by cyanogen bromide (Gross 1967). Two points can be made about these results. First, the peptide patterns of α-tubulins from the three Antarctic fish are nearly identical to each other, but differ significantly from the maps of α-chains from the temperate fish and the mammal, which, in turn, are strikingly similar. Peptides shared by digests of the Antarctic fish α-tubulins are absent, or reduced in quantity, in the maps of the dogfish and bovine α-chains, and vice versa. Second, the peptide patterns of β-tubulins from all five organisms are, in general, similar. These conclusions are supported strongly by peptide maps generated by digestion of the α- and β-tubulins by *Staphylococcus aureus* V8 protease (not shown). Thus, the α-tubulins of the cold-adapted Antarctic fish appear to have diverged structurally from those of temperate fish and mammals, whereas the β-tubulins may be more strongly conserved across these taxa.

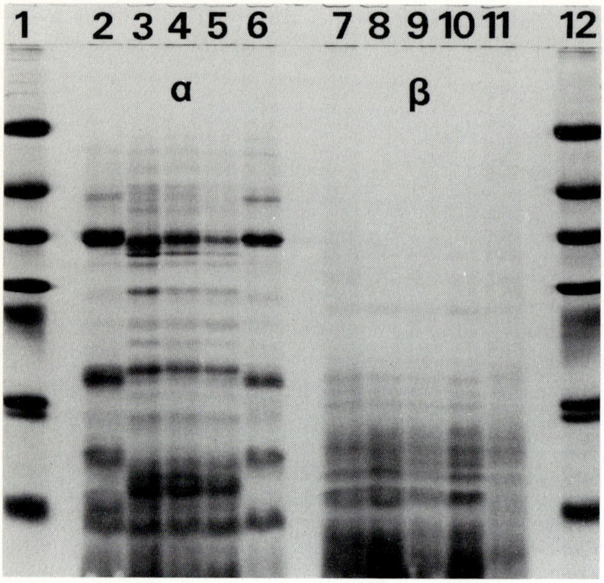

Fig. 7. Peptide maps of α- and β-tubulins from Antarctic fish, from a temperate fish, and from a mammal. Purified tubulin subunits were digested by cyanogen bromide, and the cleaved samples were electrophoresed in the presence of SDS on a 12.5% polyacrylamide gel (Detrich et al. 1987). The figure presents the peptide patterns of α- (*lanes 2–6*) and β-(*lanes 7–11*) tubulins from the dogfish shark (*lanes 2,7*), from *C. aceratus* (*lanes 3,8*), from *N. gibberifrons* (*lanes 4,9*), from *N. coriiceps neglecta* (*lanes 5,10*), and from the cow (*lanes 6,11*). *Lanes 1* and *12* contain low-molecular-weight standards, including bovine serum albumin (M_r = 66 000), ovalbumin (M_r = 45 000), glyceraldehyde-3-phosphate dehydrogenase (M_r = 36 000), carbonic anhydrase (M_r = 29 000), trypsinogen (M_r = 24 000), soybean trypsin inhibitor (M_r = 20 100), and α-lactalbumin (M_r = 14 200). (Reprinted from Detrich et al. 1987 with permission. Copyright 1987 American Society for Biochemistry and Molecular Biology)

4.4 Structure of the Carboxy-Terminal Domains of the α- and β-Tubulins

The highly acidic carboxy-terminal tails of mammalian α- and β-tubulins hinder microtubule assembly in vitro, probably because they generate electrostatic repulsion between tubulin dimers (Serrano et al. 1984; Bhattacharyya et al. 1985; Sackett et al. 1985). Thus, Detrich and Overton (1986, 1988) proposed that the tubulins of Antarctic fish polymerize efficiently at low temperatures in part because their carboxy-terminal tails have been modified, through mutation and natural selection, to contain reduced numbers of acidic residues. To compare the structures of their carboxy-terminal domains, we (Detrich et al. 1987) subjected native, colchicine-complexed tubulins from *N. coriiceps neglecta* and from the cow to limited proteolytic cleavage by subtilisin. (Colchicine was used to maintain each tubulin in its dimeric state during incubation with subtilisin at 30 °C.) We found that the piscine α-tubulin is more resistant to carboxy-terminal proteolysis than is the mammalian α-chain, whereas β-tubulins from both species are cleaved at comparable rates. Evidently, native α-tubulins from the Antarctic fish and the mammal are not structurally equivalent in their carboxy-terminal regions.

4.5 Summary of Structural Studies

The preceding results suggest, but do not prove, that many of the adaptations that enable the tubulins of Antarctic fish to polymerize at low temperatures may reside in their unique α-chains. Thus, our current efforts are directed toward determination of the primary structures and posttranslational modifications of both the α- and the β-tubulins of *N. coriiceps neglecta*.

5 Discussion

5.1 Temperature Adaptation of Microtubule Assembly

Temperature plays a crucial role in the assembly of microtubules from their subunit proteins, tubulin $\alpha\beta$-dimers and MAPs. Within temperature ranges consistent with subunit stability, microtubule formation is favored by high temperatures, while depolymerization is promoted by low temperatures. Williams et al. (1985) have proposed that cold-living poikilotherms have overcome the destabilizing effects of low temperatures on microtubules through the evolution of modified tubulins that possess appropriately large association constants. Our results provide strong support for this interpretation. We find that pure brain tubulins from three species of Antarctic fish polymerize well at temperatures as low as 0 °C. Special MAPs or dissociable, cold-stabilizing ligands are not required for polymerization of the fish tubulins at physiological temperatures. Thus, with respect to polymerization at low temperatures, the primary locus of functional adaptation appears to be the tubulin dimer.

5.2 Energetics of Microtubule Assembly by Tubulins from Poikilotherms and Homeotherms

The large, positive values for the standard enthalpy change ($\Delta H^\circ = +26.9$ kcal/mol) and the standard entropy change ($\Delta S^\circ = +123$ eu) indicate that the polymerization of Antarctic fish tubulins is entropically driven. Williams et al. (1985) reported qualitatively similar results for tubulin from another Antarctic fish, *Pagothenia borchgrevinki*, but their values for ΔH° and ΔS° (+13.7 kcal/mol and +74 eu, respectively) are somewhat smaller than our own. [This discrepancy probably reflects methodological differences rather than species identity; the reader is referred to Detrich et al. (1989) for a complete discussion.] By contrast, the enthalpies and entropies of polymerization for pure tubulins from temperate poikilotherms or from homeotherms are considerably smaller (Detrich et al. 1989). For example, Robinson and Engelborghs (1982) found that the van't Hoff plot for the polymerization of porcine brain tubulin (in a buffer system containing dimethyl sulfoxide) between 10 and 35 °C was linear and gave $\Delta H^\circ = +6.3$ kcal/mol, $\Delta S^\circ = +44.4$ eu. From published data (Detrich and Wilson 1983; Suprenant and Rebhun 1983; Detrich et al. 1985), we estimate that the standard enthalpy and entropy changes for polymerization of egg tubulin from the sea urchin *Strongylocentrotus purpuratus* (physiological temperature range = 15–18 °C) are ca. 11–16 kcal/mol and ca. 61–79 eu, respectively. Although the buffer conditions used in these studies were not identical, it appears that the entropic control over microtubule assembly increases as average body temperature decreases.

The interspecific differences in polymerization energetics apparently have an important adaptive consequence: the conservation of the critical concentration for microtubule assembly. For example, the critical concentrations for polymerization of tubulins from cold-adapted poikilotherms (e.g., Antarctic fish), from temperate poikilotherms (e.g., sea urchins, clams), and from homeotherms (e.g., mammals), when measured in vitro in buffers of similar composition and at the physiological body temperatures of these organisms, fall within the range 0.36–2.5 mg/ml (Herzog and Weber 1977; Suprenant and Rebhun 1983, 1984; Detrich et al. 1985; Williams et al. 1985). Thus, organisms from disparate thermal niches are able to assemble microtubules effectively at their normal body temperatures.

5.3 Molecular Adaptations of Antarctic Fish Tubulins

At any given temperature, the strength of the interaction between the tubulin dimers of Antarctic fish, measured by the association constant for microtubule elongation, is substantially greater than that for mammalian tubulins. The enhanced polymerization capacity of the Antarctic fish tubulins must result from changes in the molecular interactions between tubulin dimers in the polymer. Presumably, polymer-stabilizing interactions between dimers may be increased in strength, destabilizing interactions may be reduced, or changes of both types may be involved. The results of our studies are most consistent with the last possibility. Before considering the nature and location of the adaptations of the fish tubulins, a brief review of the structural domains of the tubulin subunits of mammals will be presented.

Each of the monomers of the native tubulin $\alpha\beta$-heterodimer appears to be folded to give a bilobed structure composed of larger amino-terminal and smaller carboxy-terminal globular domains (roughly two-thirds and one-third of each subunit, respectively) linked by an exposed connecting region (Mandelkow et al. 1985; Sackett and Wolff 1986). In addition, a short, extended, highly acidic carboxy-terminal tail terminates each chain (Sackett and Wolff 1986). [Each subunit may actually be composed of three subdomains of similar size (de la Viña et al. 1988), but in this discussion we will use the two-domain model.] The major *intra*dimer contacts appear to be formed by the amino-terminal domain of α-tubulin and the carboxy-terminal domain of the β-chain (Kirchner and Mandelkow 1985; Serrano and Avila 1985). Conversely, the *inter*dimer interaction within the microtubule should be formed by the carboxy-terminal domain of α-tubulin and the amino-terminal domain of the β-chain (Kirchner and Mandelkow 1985). The highly charged carboxy-terminal tails of both subunits project into the solvent and make few contacts with other tubulin domains, both in the free dimer and in the microtubule (Breitling and Little 1986; Sackett and Wolff 1986).

Two lines of evidence suggest that the functional adaptation of the Antarctic fish tubulins is based, at least in part, on an increased reliance on hydrophobic contacts between tubulin dimers. First, the polymerization energetics ($\Delta H^\circ > 0$, $\Delta S^\circ > 0$) indicate that entropy-generating interactions, (i.e., hydrophobic interactions and/or electrostatic bonds) mediate the association. In fact, the contribution of entropy-driven interactions to microtubule assembly appears to be the greatest for the Antarctic fish tubulins (see preceding section). Second, the minimal perturbation of the critical concentration (thus, the association constant for elongation) by high ionic strength demonstrates that the increased entropic control results from an increased dependence on hydrophobic interactions (relative to ionic bonds). Given the destabilizing influence of low temperatures on entropically driven reactions, this conclusion may seem paradoxical. Evidently, Antarctic fish have overcome this problem through the evolution of tubulin dimers that form increased numbers of, and/or qualitatively stronger, hydrophobic interactions at their interdimer contact surfaces. Broadly speaking, we would anticipate that changes of this type would be located primarily in the carboxy-terminal (small) globular domain of the α-chain and/or the amino-terminal (large) globular domain of β-tubulin.

A reduction in repulsive interactions produced by the negatively charged carboxy-terminal tails of the tubulin chains would also be expected to strengthen the interaction between tubulin dimers. Circumstantial evidence suggests that the Antarctic fish tubulins may have fewer negative charges in their carboxy-terminal tails than do tubulins from other organisms. In comparison to bovine tubulins, the tubulins of Antarctic fish possess isoelectric points that are generally more basic, contain fewer glutamyl and/or glutaminyl residues, and display fewer net negative charges in their native conformations (Detrich and Overton 1986, 1988). Furthermore, the carboxy-terminal tail of *N. coriiceps neglecta* α-tubulin (in the native tubulin dimer) is relatively resistant to proteolysis by subtilisin under conditions conducive to carboxy-terminal cleavage of bovine α-tubulin. Together, these results suggest that carboxy-terminal charge reduction may contribute to the enhanced polymerization efficiency of the Antarctic fish tubulins, but a rigorous test of this hypothesis will require further structural and functional studies.

5.4 Subunit Specificity of Temperature Adaptations

The interspecific conservation of the critical concentration for microtubule assembly argues strongly for the hypothesis that tubulins can acquire, through mutation and natural selection, modifications that optimize polymerization under different thermal regimes. For Antarctic fish, many, if not most, of the adaptations that favor tubulin polymerization at low temperatures are likely to be located in their structurally divergent α-tubulins. Given the functional constraints on the tubulin subunits and the recent evolutionary divergence of Antarctic fish from temperate fish, we consider it unlikely that the α-tubulins of Antarctic fish have diverged solely through the accumulation of selectively neutral alterations in their primary sequences and/or posttranslational modifications. In contrast to the α-tubulins, the β-tubulins are strikingly similar to those of temperate fish and mammals. Thus, we suggest that adaptation of microtubule assembly to low temperatures may be achieved largely through structural alterations of the α-tubulin subunits.

Acknowledgments. This paper is based on work supported by National Science Foundation Grants DPP-8317724 and DPP-8614788. I am deeply indebted to Drs. R.H. Himes, K.A. Johnson, R.F. Ludueña, and S.P. Marchese-Ragona, and to Ms. V. Prasad for their contributions to this work and for many valuable discussions and suggestions. I also wish to thank Dr. Ludueña for his permission to present Fig. 6. Finally, I gratefully acknowledge the logistical support provided to the project by the staff of the Division of Polar Programs of the National Science Foundation, by the personnel of ITT Antarctic Services, Inc., and by the captains and crews of *R/V Polar Duke.*

References

Bhattacharyya B, Sackett DL, Wolff J (1985) Tubulin, hybrid dimers, and tubulin S: stepwise charge reduction and polymerization. J Biol Chem 260:10208–10216

Breitling F, Little M (1986) Carboxy-terminal regions on the surface of tubulin and microtubules: epitope locations of YOL1/34, DM1A and DM1B. J Mol Biol 189:367–370

Correia JJ, Williams RC Jr (1983) Mechanisms of assembly and disassembly of microtubules. Annu Rev Biophys Bioeng 12:211–235

Crestfield AM, Moore S, Stein WH (1963) The preparation and enzymatic hydrolysis of reduced and S-carboxymethylated proteins. J Biol Chem 238:622–627

Detrich HW III, Overton SA (1986) Heterogeneity and structure of brain tubulins from cold-adapted Antarctic fishes: comparison to brain tubulins from a temperate fish and a mammal. J Biol Chem 261:10922–10930

Detrich HW III, Overton SA (1988) Antarctic fish tubulins: heterogeneity, structure, amino acid compositions, and charge. Comp Biochem Physiol 90B:593–600

Detrich HW III, Wilson L (1983) Purification, characterization, and assembly properties of tubulin from unfertilized eggs of the sea urchin *Strongylocentrotus purpuratus.* Biochemistry 22:2453–2462

Detrich HW III, Jordan MA, Wilson L, Williams RC Jr (1985) Mechanism of microtubule assembly: changes in polymer structure and organization during assembly of sea urchin egg tubulin. J Biol Chem 260:9479–9490

Detrich HW III, Prasad V, Ludueña RF (1987) Cold-stable microtubules from Antarctic fishes contain unique α-tubulins. J Biol Chem 262:8360–8366

Detrich HW III, Johnson KA, Marchese-Ragona SP (1989) Polymerization of jakntarctic fish tubulins at low temperatures: energetic aspects. Biochemistry 28:10085–10093

DeWitt HH (1971) Coastal and deep-water benthic fishes of the Antarctic. In: Bushnell VC (ed) Antarctic map folio series, folio 15. American Geographical Society, New York, pp 1–10

Dustin P (1984) Microtubules, 2nd edn. Springer, Berlin Heidelberg New York Tokyo

Gaskin F, Cantor CR, Shelanski ML (1974) Turbidimetric studies of the in vitro assembly and disassembly of porcine neurotubules. J Mol Biol 89:737–758

Gross E (1967) The cyanogen bromide reaction. Methods Enzymol 11:238–255

Herzog W, Weber K (1977) In vitro assembly of pure tubulin into microtubules in the absence of microtubule-associated proteins and glycerol. Proc Natl Acad Sci USA 74:1860–1864

Himes RH, Detrich HW III (1989) Dynamics of Antarctic fish microtubules at low temperatures. Biochemistry 28:5089–5095

Himes RH, Burton PR, Gaito JM (1977) Dimethyl sulfoxide-induced self-assembly of tubulin lacking associated proteins. J Biol Chem 252:6222–6228

Johnson KA, Borisy GG (1975) The equilibrium assembly of microtubules in vitro. In: Inoue' S, Stephens RE (eds) Molecules and cell movement. Raven, New York, pp 119–139

Kirchner K, Mandelkow E-M (1985) Tubulin domains responsible for assembly of dimers and protofilaments. EMBO J 4:2397–2402

Lee JC, Timasheff SN (1975) The reconstitution of microtubules from purified calf brain tubulin. Biochemistry 14:5183–5187

Lee JC, Timasheff SN (1977) In vitro reconstitution of calf brain microtubules: effects of solution variables. Biochemistry 16:1754–1764

Lowry JK (1975) Soft bottom macrobenthic community of Arthur Harbor, Antarctica. In: Pawson DL (ed) Biology of the Antarctic seas V, Antarctic Research Series 23. American Geophysical Union, Washington, pp 1–19

Mandelkow E-M, Herrmann M, Rühl U (1985) Tubulin domains probed by limited proteolysis and subunit-specific antibodies. J Mol Biol 185:311–327

Olmsted JB, Borisy GG (1975) Ionic and nucleotide requirements for microtubule polymerization in vitro. Biochemistry 14:2996–3005

Roach MC, Ludueña RF (1984) Different effects of tubulin ligands on the intrachain cross-linking of β_1-tubulin. J Biol Chem 259:12063–12071

Robinson J, Engelborghs Y (1982) Tubulin polymerization in dimethyl sulfoxide. J Biol Chem 257:5367–5371

Ross PD, Subramanian S (1981) Thermodynamics of protein association reactions: forces contributing to stability. Biochemistry 20:3096–3102

Sackett DL, Wolff J (1986) Proteolysis of tubulin and the substructure of the tubulin dimer. J Biol Chem 261:9070–9076

Sackett DL, Bhattacharyya B, Wolff J (1985) Tubulin subunit carboxyl termini determine polymerization efficiency. J Biol Chem 260:43–45

Serrano L, Avila J (1985) The interaction between subunits in the tubulin dimer. Biochem J 230:551–556

Serrano L, de la Torre J, Maccioni RB, Avila J (1984) Involvement of the carboxy-terminal domain of tubulin in the regulation of its assembly. Proc Natl Acad Sci USA 81:5989–5993

Suprenant KA, Rubhun LI (1983) Assembly of unfertilized sea urchin egg tubulin at physiological temperatures. J Biol Chem 258:4518–4525

Suprenant KA, Rubhun LI (1984) Purification and characterization of oocyte cytoplasmic tubulin and meiotic spindle tubulin of the surf clam, *Spisula solidissima*. J Cell Biol 98:253–266

de la Viña S, Andreu D, Medrano FJ, Nieto JM, Andreu JM (1988) Tubulin structure probed with antibodies to synthetic peptides. Mapping of three major types of limited proteolysis fragments. Biochemistry 27:5352–5365

Williams RC Jr, Correia JJ, DeVries AL (1985) Formation of microtubules at low temperatures by tubulin from Antarctic fish. Biochemistry 24:2790–2798

Life in Arctic Environments: Molecular Adaptation of Oxygen-Carrying Proteins

B. Giardina[1], S.G. Condò[1], A. Bardgard[2], and O. Brix[2]

1 Introduction

The metabolic needs of peripheral tissues are such that organisms have been forced, in the evolutionary sense, to develop special systems to transport oxygen from the outer environment to the tissues. The chemical basis for oxygen transport is represented by the so-called respiratory proteins, namely, hemoglobins, hemocyanins, and hemerythrins, which differ greatly in the nature of the prosthetic group and of the proteins moiety (Brunori et al. 1982; Brunori et al. 1985). In order to ensure an adequate supply of oxygen to all parts of the organism in which they occur, these proteins have developed, in the course of evolution, a common molecular mechanism based on the principle of ligand-linked conformational change in a multi-subunit structure (Wyman 1968; Perutz 1970; Antonini and Brunori 1971).

Within the framework of this common mechanism, however, different oxygen-carrying proteins have acquired special features to meet special needs. This is very clearly illustrated by certain hemoglobins from Arctic mammals and by the hemocyanin from the krill *Meganyctiphanes norvegica*, which has recently been investigated. The object of this work is to present the data obtained on these oxygen-binding systems. They are of great interest not only because they illustrate the variations possible within the scope of an overall allosteric mechanism but, even more, because these proteins represent a type case of molecular adaptation to the different physiological requirements related to life in cold environments. In fact, although temperature regulation has been studied in Arctic mammals for the last century, no study has to our knowledge investigated the role of respiratory pigments in these animals with respect to unloading of oxygen in tissues which are colder than the deep core temperature. The results obtained outline the presence of sophisticated control mechanisms, which in some cases involve the interplay of temperature and heterotropic ligands such as organic phosphates and carbon dioxide.

[1] Department of Experimental Medicine and Biochemical Sciences, University of Rome, Tor Vergata, via Orazio Raimondo, 00173 Rome, Italy.
[2] Zoological Laboratory, University of Bergen, Allegaten 41, 5007 Bergen, Norway.

Guido di Prisco (Ed)
Life Under Extreme Conditions

2 Materials and Methods

The reindeer (*Rangifer tarandus tarandus*) and the musk-ox (*Ovibos muschatos*) had been bred out of doors by the Department of Zoology, University of Oulu, Finland, and by the Department of Arctic Biology, University of Tromso, Norway. The cervus (*Cervus elaphus*) was caught near Tromso, Norway. Two specimens of the Lesser Rorqual, *Balaenoptera acutorostrata*, were caught on the Norwegian Scientific Whale Cruise 1988 arranged by the Biological Institute, Oslo, Norway.

The blood was sampled in heparinized syringes and stored in ice water. The cells were washed three times with isotonic NaCl solution and the packed cells lysed by adding 2 vol of cold distilled water. The stroma were removed by centrifugation at 10 000 rpm for 30 min. Electrophoretic analysis was performed by alkaline polyacrylamide gel electrophoresis. Stripped hemoglobin was obtained by passing the hemolysate over a mixed-bed ion exchange column (Dowex AG 501×8).

Meganyctiphanes norvegica was collected during a 4 day cruise in the North Sea (approximately, 60 °10′ N 3 °20′ E) in the middle of August 1984.

The blood samples were pooled from 766 specimens (mean size 277–40 mm) into Eppendorf tubes and frozen. Prior to experiments, the frozen samples were thawed and centrifuged to remove gelled blood. The blood was stored in iced water until use.

Concentrated stock solutions of 2,3-DPG were prepared by dissolving the sodium salt (Sigma) in water or in buffer solutions.

Concentrated stock solutions of P6-inositol (0.1 M) were prepared by dissolving the sodium salt of phytic acid (Sigma) in water and adjusting the pH to the desired value with concentrated phosphoric acid.

Oxygen dissociation curves were determined spectrophotometrically with the tonometric method (Giardina and Amiconi 1981) at a protein concentration of 3–5 mg/ml. Alternatively oxygen dissociation curves were obtained with a diffusion chamber technique (Lykkeboe et al. 1975) except that the stepwise increases in oxygen tension were generated by cascaded gas mixing pumps (Wosthoff) while absorbance changes between zero and full saturation were monitored on recorders with high sensitivity units (Eppendorf, Radiometer). The application of small amounts of hemoglobin solution (layer thickness about 10 μm) minimized equilibration solution time and methemoglobin formation, which was always $< 1\%$. Spectrophotometric measurements were carried out with a Perkin-Elmer 550 SE spectrophotometer.

The oxygen-binding data were fitted to the MCW model (Monod et al. 1965) as previously described (Brix et al. 1989a). The change in enthalpy, δH, accompanying oxygenation was calculated from the integrated van't Hoff equation:

$$\delta H = -4{,}574\ (T_1 T_2 / T_1 - T_2)\ \delta \log P_{50} / 1000\ \text{kcal mol.}$$

The pH was measured on oxygenated and deoxygenated blood under the same conditions as the oxygen equilibrium curves. The mean value has been used in the presentation of the data.

3 Results

3.1 Hemoglobin Properties of Reindeer (*Rangifer tarandus tarandus*), Musk-Ox (*Ovibos muschatos*), and Cervus (*Cervus elaphus*)

These animals live in sub-Arctic areas beyond latitude 65° and are exposed to extremes of a variety of environmental factors including nutritional, climatic, and light rhythms. We studied in detail the functional properties of their hemoglobins with special regard to the effect of temperature, since it is known that they encounter a great range of temperature changes from –40 to +20 °C.

As far as the general properties of these hemoglobins are concerned, their reactions with oxygen and protons are functionally linked and qualitatively similar to those found for other mammalian hemoglobins. Moreover, all the three proteins appear to belong to the group whose oxygen affinity is not modulated in vivo through the interactions with 2,3-DPG (Bunn 1980; Perutz and Imai 1980). In other words, these hemoglobins, in the presence of chloride ions at physiological concentration, are almost insensitive to the presence of this effector.

The peculiar feature which characterizes all of them is the minor effect of temperature on the oxygen binding. It should be recalled, in fact, that the overall heat required in the oxygenation of mammalian hemoglobins ranges, in general, from –13.5 to –14.5 kcal/mol of oxygen, which, after correction for the heat of oxygen solubilization (–3.0 kcal/mol), gives values from –10.5 to –11.5 kcal/mol of oxygen (Antonini and Brunori 1971). However, in the case of reindeer, musk-ox, and cervus under physiological conditions, the temperature sensitivity expressed by the overall heat of oxygenation, δH, was found to be almost three times lower than that of human hemoglobin under the same experimental conditions (Condò et al. 1988; Brix et al. 1989a; Giardina et al. 1989a). In the case of Hb from musk-ox, this is evident from the data reported in Fig. 1, where the Bohr effect obtained at two different temperatures has been reported. It should be mentioned here that the same type of results are obtained in the presence of 3 mM 2,3-DPG, confirming the inefficacy of this molecule in the modulation of musk-ox hemoglobin. The apparent δH values obtained indirectly from the van't Hoff equation are reported in the inset of Fig. 1. In the case of human hemoglobin A going toward acid pH values, the apparent δH of oxygen binding becomes less and less exothermic due to the increasing contribution of the Bohr protons that cancels some of the heat released upon oxygen binding. However, in the case of musk-ox, the apparent heat of oxygenation is at its maximum value, even if small, just within the physiological pH range. It tends to zero or even positive values, going both toward more acid and more alkaline pH values. Hence, these very small or even positive δH values are obtained in regions of pHs where the alkaline Bohr effect is disappearing. We may therefore exclude a significant involvement of the Bohr protons in determining this unusual δH of oxygen binding. We may consider either an intrinsic property of the molecule or the effect of some other ions, whose presence could be important in vivo in determining the overall functional property of the hemoglobin molecule.

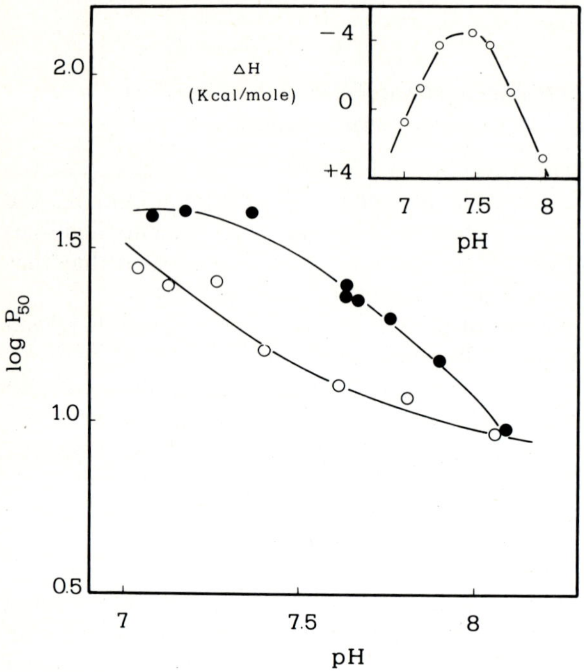

Fig. 1. Oxygen Bohr effect of musk-ox hemoglobin at 20 °C (*open circles*) and 37 °C (*closed circles*) in 0.1 M bistris or Tris buffer plus 100 mM NaCl. The *inset* reports the apparent heat of oxygenation, calculated from the integrated van't Hoff equation, as a function of pH. The values are corrected for the heat contribution of oxygen in solution (After Brix et al. 1989a)

That the small δH of oxygen binding could be a general characteristic of the hemoglobin from Arctic mammals is further indicated by the data obtained on hemoglobin from cervus (Fig. 2). Here, the overall heat of oxygen binding is very small, positive at physiological pH values, and displays a behavior opposite to human hemoglobin A, as far as the pH dependence is concerned. Thus, as the pH is brought toward acid values, the overall heat of oxygenation becomes more and more exothermic, indicating the presence of structural changes which overcome the endothermic contribution of the Bohr protons.

The limited effect of temperature on the oxygen binding in reindeer hemoglobin has been investigated in great detail by a set of experiments carried out as a function of temperature (Giardina et al. 1989b). Typical results are shown in Fig. 3. The data, presented in the form of a Hill plot, extend over a saturation range broad enough to permit an evaluation of a number of thermodynamic parameters. One feature which shows up very clearly is the strong temperature dependence of the shape of the binding curve. Thus, an increase of temperature brings about a great decrease in the association constant for the binding of the first oxygen molecule, without significantly affecting that for the binding of the last

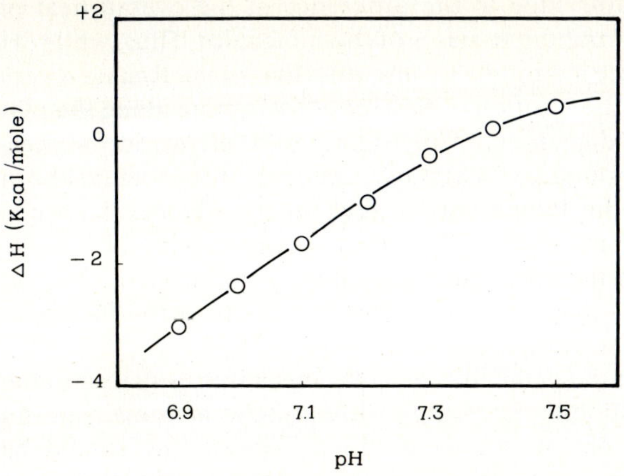

Fig. 2. Apparent heat of oxygenation (ΔH) values for cervus elaphus hemoglobin within the physiological pH range. Values were obtained from oxygen equilibrium experiments performed in 0.1 M Tris buffer plus 100 mM NaCl at 20 and 37 °C and are corrected for the heat contribution of oxygen in solution

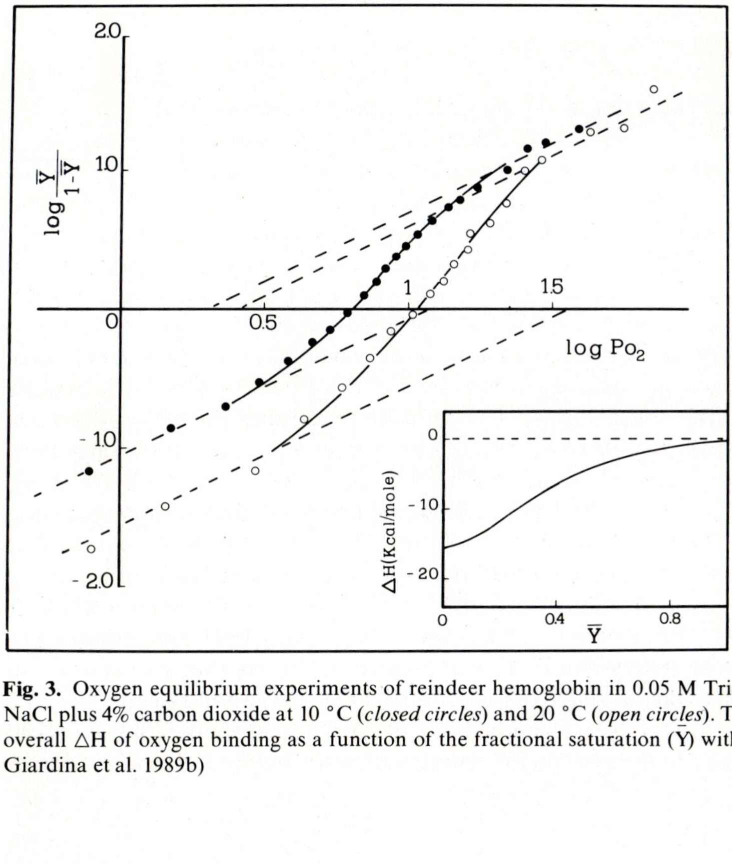

Fig. 3. Oxygen equilibrium experiments of reindeer hemoglobin in 0.05 M Tris buffer plus 0.1 M NaCl plus 4% carbon dioxide at 10 °C (*closed circles*) and 20 °C (*open circles*). The *inset* reports the overall ΔH of oxygen binding as a function of the fractional saturation ($\bar{Y}$) with this ligand (After Giardina et al. 1989b)

oxygen molecule. This is mainly due to the difference in the overall heat of oxygenation relative to the T- and the R-states of the molecules. Thus, while δH of oxygen binding to the T-state is strongly exothermic, that of the R-state is very close to zero. In other words, the position of the upper asymptote along the pO_2 axis is almost temperature-independent. This dramatic difference in the thermodynamics of the two conformational states of reindeer hemoglobin results in a particular dependence of the temperature effect on the degree of oxygen saturation of the protein (see inset to Fig. 3). As a consequence, within the range of oxygen saturation in which the protein is working in vivo, the overall heat of oxygenation is of the order of –3.8 kcal/mol, approaching zero as the fractional saturation tends to 1.

In conclusion, apart from the detailed molecular mechanisms that underlie this phenomenon, the hemoglobins from these three Arctic mammals are all characterized by very small δH of oxygen binding. This finding should be considered in connection with the very low environmental temperature that these animals are required to face in their living habitat within the year. On the whole, these proteins may be regarded as a beautiful example of molecular adaptation to extreme environmental conditions. Thus, they do not need much energy during their oxygenation-deoxygenation cycle, and oxygen delivery is not drastically impaired at the level of peripheral tissues such as skin, legs, etc. where temperature may be significant lower (up to about 10 °C) than that of the lungs (Giardina et al. 1989b).

3.2 Hemoglobin Properties of Whale (*Balaenoptera acutorostrata*)

The functional properties of the hemoglobin from the whale have been characterized as a function of pH, organic phosphates, carbon dioxide, and temperature. As evident from Fig. 4, whale hemoglobin is characterized at 20 °C by an oxygen affinity which is significantly higher (about 30%) than that of human hemoglobin A, when compared in the absence of organic phosphates and carbon dioxide. As expected when carbon dioxide is added to the protein solution, a pronounced decrease in oxygen affinity is observable. A marked decrease of oxygen affinity is also observed when the molecule is saturated with organic phosphates (P6-inositol in Fig. 4). It should be noted that the carbon dioxide-induced decrease of oxygen affinity is present even in the presence of a saturating concentration of organic phosphates (Fig. 4).

The data obtained at 37 °C and reported in the same figure outline three main characteristics: (1) under stripped conditions, i.e., in the absence of organic phosphates and carbon dioxide, an increase in temperature brings about a great decrease in the oxygen affinity; (2) at 37 °C, the effect of carbon dioxide is completely abolished, implying the disappearance of the preferential binding of this effector to the deoxy-conformational state of the protein (Antonini and Brunori 1971; Kilmartin and Rossi-Bernardi 1973); (3) although the intrinsic temperature sensitivity is extremely high (Table 1), when the physiological cofactors are added to the system the overall heat required for the oxygenation of

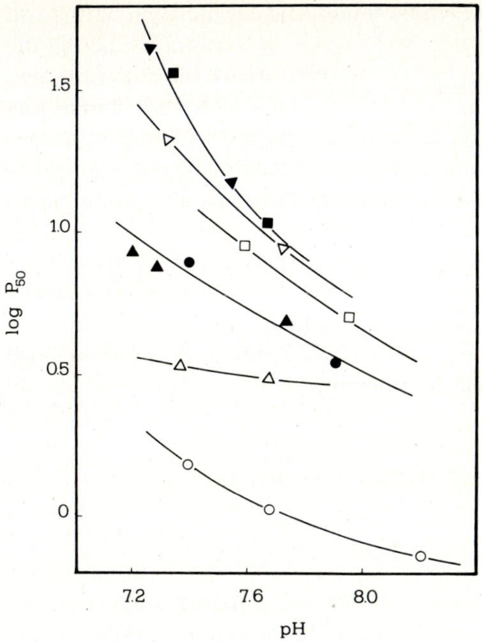

Fig. 4. Oxygen Bohr effect of *Balaenoptera acutorostrata* hemoglobin in 0.1 M Tris buffer plus 100 mM NaCl at 20 °C (*open symbols*) and 37 °C (*closed symbols*). Stripped hemoglobin (*circles*); plus 2% carbon dioxide (*triangles*); stripped Hb plus 30 mM P6-inositol (*squares*); stripped Hb plus 30 mM P6-inositol plus 2% carbon dioxide (*reversed triangles*)

Table 1. Heat of oxygenation (δH) in kcal/mol of whale hemoglobin (in 0.1 M Tris buffer plus 0.1 M NaCl at pH 7.4)

Stripped Hb	−16.2
Plus 3 mM 2,3-DPG	− 4.6
Plus 30 mM P6-inositol (no CO_2)	− 6.1
Plus 30 mM P6-inositol (saturated with CO_2)	− 2.0

whale Hb, after correction for the heat of oxygen solubilization, ranges from −4 to −2 kcal/mol of oxygen (Table 1).

This latter feature brings the whale hemoglobin into the same category of hemoglobins from Arctic mammals such as reindeer, musk-ox, and cervus (Brix et al. 1989c). The significance of such a low temperature dependence of oxygen binding in hemoglobin of whale may appear if we consider that, although the major part of the whale's body is covered by a thick isolating layer of blubber, the metabolic very active parts, i.e., the flippers and the large tail, are not so well insulated. They are kept at a lower temperature by a countercurrent heat exchangers in order to reduce heat loss (Scholander and Schevill 1955). The case of oxygen unloading in these active regions of the whale's body is thus very similar to that of oxygen unloading in the cold leg muscles of the reindeer, musk-ox, and cervus (Brix et al. 1989a; Giardina et al. 1989b). The major difference between whale and terrestrial mammals lies in the molecular mechanisms of achieving such a low temperature sensitivity.

Finally, the lower temperature that whale blood finds at the level of fins and tail, in comparison with the rest of the body, gives a rationale also to the temperature dependence of the effect of carbon dioxide. Thus, the data obtained indicate that, within the core of the organism, carbon dioxide does not display any allosteric effect; at 37 °C, the differential binding of this ligand with respect to oxy- and deoxy-structure is completely abolished. However, carbon dioxide returns to facilitate oxygen unloading just at the level of fins and tail, which are the places of great muscular activity and whose temperature is well below 37 °C, due to the presence of a countercurrent heat exchanger. In order to save energy, this allows temperatures to decrease at the level of these less insulated parts of the body.

In conclusion, the combined effects of organic phosphates, carbon dioxide, and temperature seen in whale hemoglobin optimize oxygen delivery at the level of all tissues in spite of the great local heterothermia.

3.3 Hemocyanin from the Krill (*Meganyctiphanes norvegica*)

This epipelagic species lives in the north Atlantic waters (Mauchlin and Fisher 1969) and is abundant throughout the year in Norwegian fjords. Its hemocyanin is characterized by a relatively low oxygen affinity and a high cooperativity of oxygen binding (Brix et al. 1989b). Moreover, for pH values higher than 7.3, the oxygen affinity increases markedly as a function of temperature; this feature reflects a strong endothermic overall heat of oxygenation (see Fig. 5) which is a

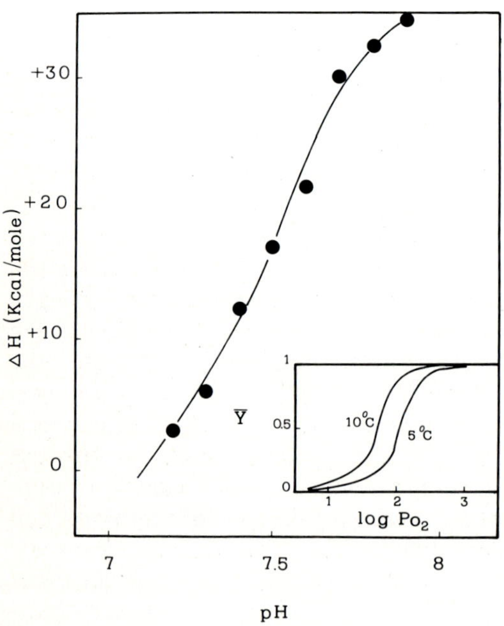

Fig. 5. pH dependence of the apparent overall heat of oxygenation of *Meganyctiphanes norvegica* hemocyanin. The *inset* reports the oxygen-binding isotherms of the hemocyanin at two different temperatures; pH = 7.7

novel characteristic never previously reported for respiratory pigments. The apparent δH of oxygen binding to krill hemocyanin decreases as the proton concentration is increased, just in the pH range in which the alkaline Bohr effect is operative. Once again this behavior is opposite to that expected on the basis of the endothermic contribution of the Bohr protons, and is in some way reminiscent of the case of cervus hemoglobin, discussed above. Hence, also in the case of krill hemocyanin, excluding any masking effect of the Bohr protons, the possible explanation for the endothermic character of oxygen binding has to be related either to an intrinsic property of the macromolecule, or to a specific effect of ions that still needs clarification.

The unusual thermal properties of this hemocyanin remain per se a phenomenon of striking adaptive and evolutionary significance, since they fit very well to the behavior of the animal which descends at daytime down to 100–400 m and ascends at night up to 100–0 m depth (Mauchlin 1980). In our opinion, the strong endothermic character of the oxygen binding should be of great benefit for the krill during its feeding excursions when it ascends to the upper layers, where the oxygen availability may be significantly reduced due to the lower solubility at higher temperature and to the photorespiratory activity of the phytoplankton (Brix et al. 1989b).

4 Conclusion

The results discussed in this chapter are of particular evolutionary interest and outline the role of temperature and of its interplay with some heterotropic ligands in the modulation of respiratory pigments. Moreover, these studies have clearly pointed out that much interesting information can be overlooked when experiments are carried out at a single temperature, not taking into account that the thermodynamic parameters of oxygen binding are of extreme importance also from a physiological point of view. In this respect, the energy-saving mechanism displayed by reindeer, musk-ox, and cervus is very instructive, and certainly one of the most elegant examples of the different strategies which have been adopted during evolution to solve the problem of oxygen transport to respiring tissues. Furthermore, the case of whale hemoglobin points out that the effect of temperature should also be studied with respect to the reaction with heterotropic effectors. Here, the interplay of temperature and carbon dioxide is particularly impressive. Thus, temperature controls the modulatory effect of carbon dioxide switching on and off the differential binding of this ligand, in dependence of the specific region of the organism.

Finally, the endothermic oxygenation of krill hemocyanin outlines that, although the different respiratory pigments provide a good example of the different chemical strategies adopted during evolution to solve the vital problem of oxygen supply for metabolic demands, the fine modulation of the basic function is achieved, at the molecular level, within the same biophysical framework. The latter is that of the allosteric control, which may in turn be finely modulated by temperature through specific thermodynamic characteristics of the macromolecule. These should lie in some structural properties which now call for future investigation.

References

Antonini E, Brunori M (1971) Hemoglobin and myoglobin in their reactions with ligands. North Holland, Amsterdam

Brix O, Bardgard A, Mathisen S, El-Sherbini S, Condò SG, Giardina B (1989a) Arctic life adaptation. II. The function of musk ox (*Ovibos muschatos*) hemoglobin. Comp Biochem Physiol 94B:135–138

Brix O, Borgund S, Barnung T, Colosimo A, Giardina B (1989b) Endothermic oxygenation of hemocyanin in the krill *Meganyctiphanes norvegica*. FEBS Lett 247:177–180

Brix O, Condò SG, Lazzarino G, Clementi ME, Scatena R, Giardina B (1989c) Arctic life adaptation. III. The function of whale (*Balaenoptera acutorostrata*) hemoglobin. Comp Biochem Physiol 94B:139–142

Brunori M, Giardina B, Kuiper HA (1982) Oxygen transport proteins. In: Hill HAO (ed) Inorganic biochemistry, vol 3. The Chemical Society, London, pp 126–182

Brunori M, Coleta M, Giardina B (1985) Oxygen carrier proteins. In: Harrison P (ed) Metalloproteins, vol 1. Macmillan, Chicago, p 263

Bunn HF (1980) Regulation of hemoglobin function in mammals. Am Zool 20:199–211

Condò SG, El-Sherbini S, Shehata YM, Serpe E, Nuutinen M, Lazzarino G, Giardina B (1988) Regulation of the oxygen affinity of hemoglobin from the reindeer (*Rangifer tarandus tarandus*). Arct Med Res 47:83–88

Giardina B, Amiconi G (1981) Measurement of binding of gaseous and nongaseous ligands to hemoglobin by conventional spectrophotometric procedures. Methods Enzymol 76:417–427

Giardina B, Condò SG, El-Sherbini S, Mathisen S, Tyler N, Nuutinen M, Bardgard A, Brix O (1989a) Arctic life adaptation. I. The function of reindeer hemoglobin. Comp Biochem Physiol 94B:129–133

Giardina B, Brix O, Nuutinen M, El-Sherbini S, Bardgard A, Lazzarino G, Condò SG (1989b) Arctic adaptation in reindeer: the energy saving of a hemoglobin. FEBS Lett 247:135–138

Kilmartin JV, Rossi-Bernardi L (1973) Interaction of hemoglobin with hydrogen ions, carbon dioxide and organic phosphates. Physiol Rev 53:836–890

Lykkeboe G, Johansen K, Maloy GM (1975) Functional properties of hemoglobin in the teleost *Tilapia grahami*. J Comp Physiol 104:1–11

Mauchlin J (1980) The biology of mysids and euphausiids. In: Blaxter JHS, Russel FS, Yonge M (eds) Advances in marine biology, vol 18. Academic Press, London

Mauchlin J, Fisher LR (1969) The biology of euphausiids. In: Russel FS, Yonge M (eds) Advances in marine biology, vol 7. Academic Press, London

Monod J, Wyman J, Changeux JP (1965) On the nature of allosteric transitions: a plausible model. J Mol Biol 12:88–118

Perutz MF (1970) Stereochemistry of cooperative effects in hemoglobin. Nature 228:726–739

Perutz MF, Imai K (1980) Regulation of oxygen affinity of mammalian hemoglobins. J Mol Biol 136:183–191

Scholander PF, Schevill WE (1955) Counter-current vascular heat exchange in the fins of whales. J Appl Physiol 8:279–282

Wyman J (1968) Regulation in macromolecules as illustrated by hemoglobin. Q Rev Biophys 1:35–80

Archaebacteria: Lipids, Membrane Structures, and Adaptation to Environmental Stresses

M. De Rosa[1,2], A. Trincone[1], B. Nicolaus[1], and A. Gambacorta[1]

1 Introduction

In the past few years a revolution has occurred in the taxonomy of living organisms. In fact, on the basis of genetic studies and on the acquisition of other general biochemical features, organisms are no longer merely gathered into two groups of eubacteria and eukaryotes, but may be considered to belong to a third line, the archaebacteria (Woese 1987).

The archaebacteria have been recognized as a phylogenetically coherent separate group of microorganisms, and differ from eubacteria and eukaryotes in a series of genetic and molecular aspects (Woese 1987). They are a collection of disparate phenotypes and thrive in environments that would normally kill many other known organisms. They are segregated into a few peculiar ecological niches, such as saturated brine for halophiles, strict anaerobiotic environment for methanogens, and thermal habitats for extreme thermophiles (Woese 1987; Konig 1988).

From the point of view of the early evolution of life, archaebacteria have to be considered quite interesting organisms, since they were probably the dominant forms of life in the primeval biosphere. Many metabolic features of these microorganisms seem to be suitable for the ecological parameters which have characterized the early history of life on earth. The importance of the archaebacteria lies in the improvement of our knowledge of the early events in the evolution of cells, contributing to a better understanding of the universal ancestor. Although archaebacterial taxonomy is complex and controversial, the new recent isolates all conform essentially to one of the three original types: halophiles, methanogens, and thermophiles (De Rosa and Gambacorta 1988).

Recent studies based on the comparison of complete 16S ribosomal RNA sequences confirm, refine, and extend earlier archaebacterial phylogeny. In particular, these studies show that archaebacteria could now be divided into the following two major branches: (1) the sulfur-dependent thermophilic organisms, and (2) the methanogens, which comprise the extreme halophiles and some

[1] Istituto per la Chimica di Molecole di Interesse Biologico CNR, Via Toiano 6, Arco Felice, Napoli, Italy.

[2] Istituto di Biochimica delle Macromolecole, Universita' di Napoli I, Facolta' di Medicina, Via Costantinopoli 16, Napoli, Italy.

Guido di Prisco (Ed)
Life Under Extreme Conditions

thermophilic species belonging to the orders of *Thermoplasmatales*, *Thermococcales* and *Archaeglobales* (Fig. 1; Woese 1987).

In view of the key role of the membrane ability of archaebacteria to survive in the face of drastic environmental stresses, extensive studies have thus been undertaken to characterize the structural identity and the biogenetic origin of their membrane lipids (Langworthy 1985; De Rosa and Gambacorta 1988; Kamekura and Kates 1988).

Upon comparing molecular components of archaebacteria with prokaryotes and eukaryotes, one is struck by the major chemical differences observed in the lipids, especially in comparison with the other components whose essential chemical features appear to be preserved. In fact, the structure of membrane lipids remains one of the major distinctions between all archaebacteria and other organisms.

All the membrane lipids of the archaebacteria so far identified are characterized by unusual structural features, which can be considered to be specific taxonomic markers of this group of microorganisms (Langworthy 1985; De Rosa and Gambacorta 1988; Kamekura and Kates 1988). Archaebacteria do not contain fatty acyl chains in their lipids; instead, they are characterized by the presence of phytanyl chains derived from saturated C_{20}, C_{25}, or C_{40} alcohols joined to glycerol or other polyols by ether linkage, resulting in ether lipids.

Moreover, it is worth noting that all glycerol ethers in archaebacteria contain a 2,3-*sn*-glycerol, which is unusual, since the glycerol, in naturally occurring glycerophosphatides or diacylglycerols, is known to have 1,2-*sn* stereochemistry.

Although isoprenoids with specific functions occur in the lipid membrane of most cells, the ether lipids dealt with here are the only such compounds that provide the major structural components of the membrane in which they occur.

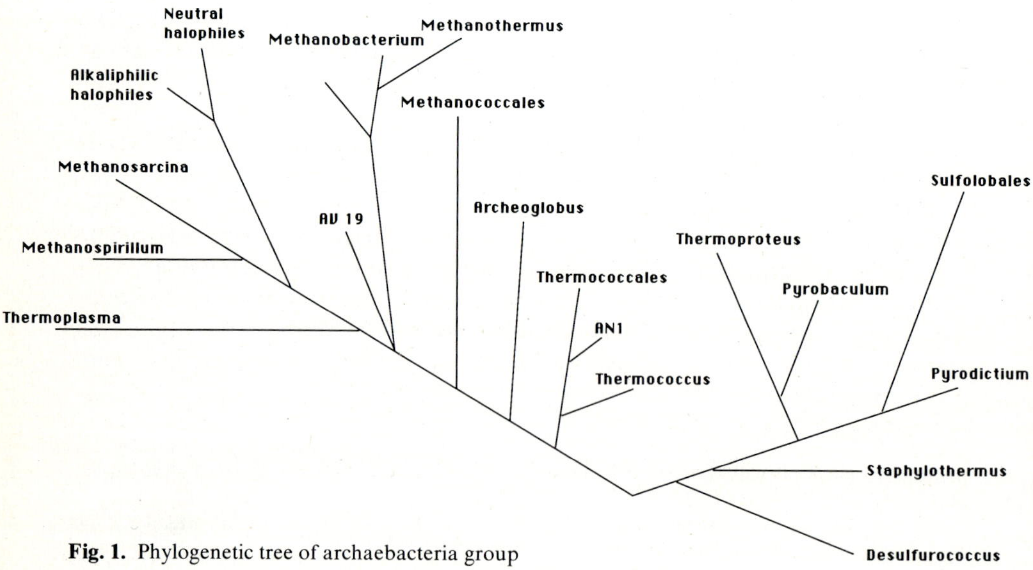

Fig. 1. Phylogenetic tree of archaebacteria group

We may wonder how archaebacterial lipids with a chemical composition so different from conventional ester lipids can manage to perform similar functions, and why that chemical variability is required for living organisms. In this respect, one of the most important aspects is the organization of ether lipids in the plasma membrane of archaebacteria.

This presentation is meant to be a survey of the lipid structure occurring in archaebacteria so far examined, and of the most important features of archaebacterial lipid biosynthesis. Moreover, some data about phenotypic adaptation of archaebacteria to environmental stresses are discussed.

2 Lipid Structure of Archaebacteria

2.1 Core Lipids

The complex lipids of archaebacteria are mainly based on two classes of isopranoid ether core lipids: diethers and tetraethers. These compounds are easily obtained by acid or alkaline hydrolysis of complex lipids (De Rosa and Gambacorta 1988).

2.2 Diether Archaebacterial Lipids

Different types of isopranoid diethers, basic structural elements of the membrane lipids occuring in different groups of archaebacteria, are reported in Fig. 2. These molecules could be formally considered the archaebacterial counterparts of the conventional diglycerides of eukaryotic or eubacterial origin.

A general feature of all archaebacterial glycerol lipids is the unusual 2,3-*sn* glycerol stereochemistry, opposite to the one found in common diglycerides, which is probably related to the biochemical mechanism of ether linkage assembly.

The diethers found in archaebacteria are reported in Fig. 2. Figure 2a shows the 2,3-di-O-phytanyl-*sn*-glycerol formed by condensation via ether linkage between two C_{20} isopranoid alcohols and a glycerol molecule. The chiral centers of the isopranoid chains have the 3R, 7R, 11R configuration. The 2,3-di-O-phytanyl-*sn*-glycerol, a (Fig. 2), could be considered a universal core lipid in such types of microorganisms. It may be present as the major component depending on the type of archaebacteria. This glycerol diether represents 100% of ether core lipids in the majority of neutral halophiles and in some coccoid forms belonging to the methanogens and thermophiles (Langworthy and Pond 1986; De Rosa and Gambacorta 1988).

Structural types b-c (Fig. 2) occur in the extremely alkaliphilic halophiles of the genera *Natronococcus* and *Natronobacterium*, living at pH 10, and in a few strains of neutral halophiles of the genera *Halobacterium* and *Halococcus*. The 2-O-sesterterpanyl-3-O-phytanyl-*sn*-glycerol, b (Fig. 1), represents up to 80% of

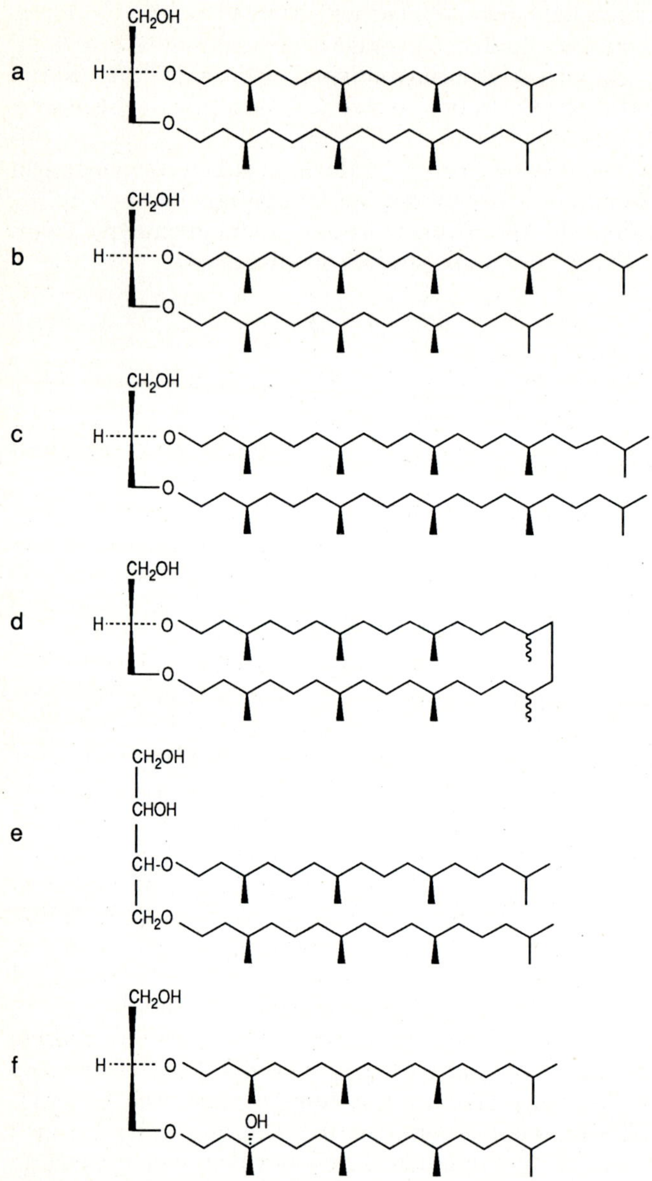

Fig. 2a-f. Isopranoid diether lipids in archaebacteria (see text)

total ether core lipids in the haloalkaliphiles (De Rosa and Gambacorta 1988; Kamekura and Kates 1988).

Macrocyclic glycerol diether, d (Fig. 2), is characterized by the presence of a 36-membered ring, which formally originates from the condensation of a glycerol moiety with a C_{40}w,w[1] isopranyl diol. From a biosynthetic point of view, the macrocyclic structure probably derives from the unprecedented head-to-head condensation of two isopranoid residues of a 2,3-di-O-geranylgeranyl-*sn*-glycerol precursor (Comita et al. 1984).

Diphytanyl tetritol diether, e (Fig. 1), isolated from *Methanosarcina barkeri* and *Methanosarcina mazei*, represents up until now the only example, among archaebacteria, of a core lipid in which glycerol is absent (De Rosa et al. 1986b; De Rosa and Gambacorta 1988).

Recently, Ferrante et al. (1988a) isolated from the aceticlastic methanogen *Methanothrix concilii* GP6 the diether core lipid, f, (Fig. 2). This new core lipid, accounting for 30% of total polar lipids of the microorganisms, is obtained only when mild hydrolytic conditions are used. It possesses, in fact, a hydroxyl group on C-3 of the phytanyl chain at the sn-3 position of the glycerol.

2.3 Tetraether Archaebacterial Lipids

Figure 3 shows two families of macrocyclic tetraethers which are the basic components of the membrane lipids of many methanogens and, with few exceptions, of all thermophilic archaebacteria (Langworthy and Pond 1986; De Rosa and Gambacorta 1988).

The tetraether a (Fig. 3) has a 72-member ring with 18 stereocenters, formed by two 2,3-*sn*-glycerol moieties, bridged through ether linkages by two isopranoid C_{40} diols formally derived from head to head linkage of two O-phytanyl residues. The series of tetraethers b-i (Fig. 3) can be considered to derive from the first one by a process of cyclization of the aliphatic component (3R, 7R, 11R, 15S, 18S, 22R, 26R, 30R)-3,7,26,30-octamethyldotriacontane, between methyls in 3,7,26,30 and methylenes 6,10,23,27, respectively. All these structures are simply named glycerol dialkyl glycerol tetraethers (GDGTs).

The second series a′-i′ (Fig. 3) has a structural organization similar to that of GDGTs, but a more complex polyol, with nine carbon atoms replacing one of the glycerols. By analogy, these compounds are named glycerol dialkyl nonitol tetraethers (GDNTs; De Rosa and Gambacorta 1988).

A high degree of structural specificity characterizes these tetraethers, originating from two polyols and five different C_{40} isopranoids with up to four cyclopentane rings per chain. In all the structures identified so far in *Sulfolobus solfataricus*, the two polyols are always antiparallel and the two C_{40} chains are either identical or differ by one cycle. Moreover, in tetraethers with an asymmetric end-to-end C_{40} chain (mono- and tricyclic C_{40} isopranoids), the more cyclized end is always linked to the primary carbinol of the polyols (De Rosa et al. 1983).

The tetraether a (Fig. 3), together with the 2,3-di-O-phytanyl-*sn*-glycerol, a (Fig. 2), in different proportions, are the basic components of complex lipids of

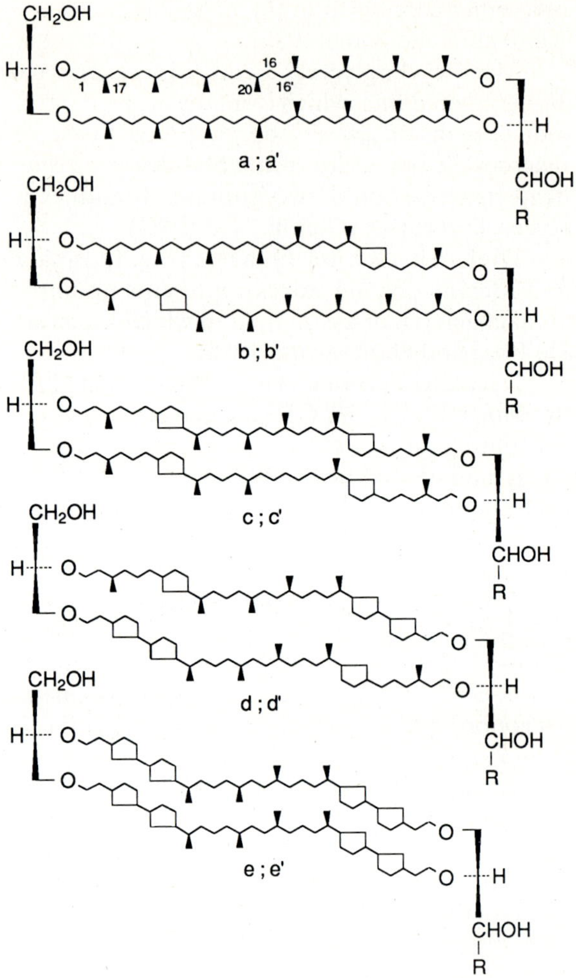

Fig. 3. Isopranoid tetraether lipids in archaebacteria. **a-i** Glycerol-dialkyl-glycerol tetraethers (GDGTs); **a′-i′**:glycerol-dialkyl-nonitol tetraethers (GDNTs) (see text)

methanogenic archaebacteria belonging to the orders *Methanobacteriales* and *Methanomicrobiales*. Only in *Methanothermus sociabilis* and in *Methanosarcina barkeri* tetraethers with cyclopentane rings in the isopranoid chains have been detected (Langworthy and Pond 1986; De Rosa and Gambacorta 1988).

GDNT lipids are found only in microorganisms belonging to the *Sulfolobales* and seem to be a specific taxonomic marker for this order (Langworthy and Pond 1986; De Rosa and Gambacorta 1988).

Tetraethers are the main components of lipids of thermophilic archaebacteria, with the exception of *Pyrococcus woesei*, *Thermococcus celer*, and AN1 isolate, whose lipids are based on diether a (Fig. 2; De Rosa et al. 1987; Lanzotti et al. 1989b).

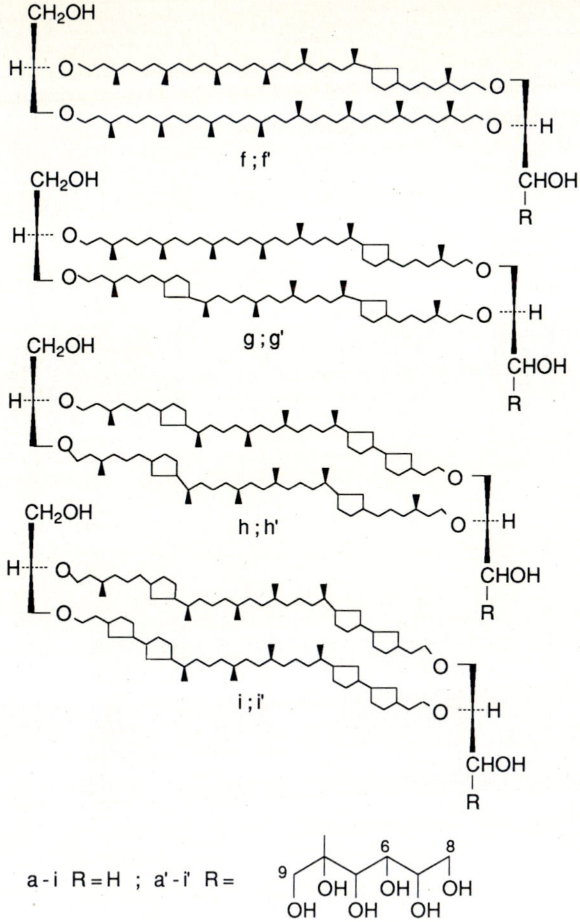

2.4 Minor Archaebacterial Lipids

In Fig. 4, three minor core lipids are reported which occur in archaebacteria.

The structure a (Fig. 4) has been isolated from lipids of *Methanosarcina barkeri* and, in trace amounts, also in *S. solfataricus* (De Rosa et al. 1986b; De Rosa et al. 1989).

Compound b (Fig. 4) is a glycerol trialkyl glycerol tetraether based upon two glycerol linked by four ether bonds with two C_{20} and one C_{40} isopranoid chain. The absence in *S. solfataricus* of tetraethers with this type of molecular architecture, but with cyclopentane rings in the isopranoid chains, as in GDGT and in GDNT (Fig. 3), could indicate that the enzyme system devoted to cyclopentane ring formation operates only if the 72-membered macrocycle is closed (De Rosa et al. 1989).

Recently, Thurl and Schafer (1988) reported the occurrence, as a minor component of tetraethers in *Thermoproteus tenax*, of a new GDGT having in

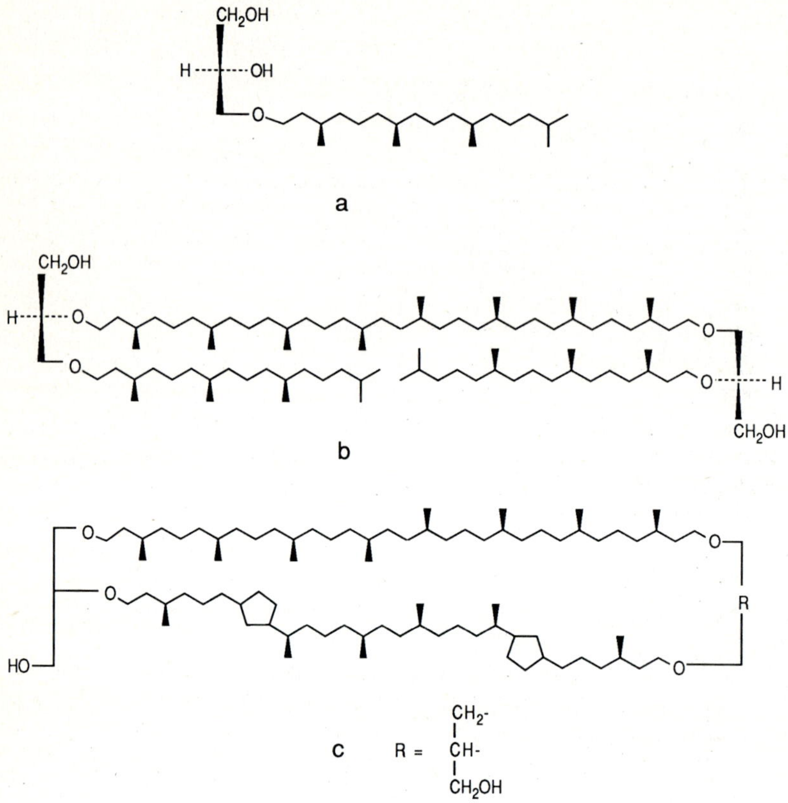

Fig. 4a-c. Minor ether lipid in archaebacteria

the macrocycle an acyclic and a bicyclic C_{40} isopranoid chain, c, (Fig. 4); the relative orientation of the two glycerol moieties in this tetraether remains unknown.

3 Polar Lipids

3.1 Halophiles

Although polar complex lipids based on 2,3-di-O-phytanyl-*sn*-glycerol are found in all archaebacterial phenotypes, they differ in the nature of the attached polar heads.

In the halophilic archaebacteria four basic complex lipid structures are found: (1) the isoprenoid ether analogues of phosphatidylglycerol (PG); (2) phosphatidylglycerophosphate (PGP); (3) phosphatidylglycerosulfate (PGS;

Fig. 5); and (4) a family of glycolipids and their sulfate derivatives, differing among themselves in the number and type of sugars (Fig. 5; De Rosa et al. 1989).

The amino lipids are absent in halophilic archaebacteria, which is unusual in that most of eubacterial counterparts usually contain phosphatidylethanolamine or lipoamino acids (Kamekura and Kates 1988).

Recently, Tsujimoto et al. (1989) and Fredrickson et al. (1989) revised the structure of PGP in *Halobacterium halobium* and *H. cutirubrum*, respectively, by using the FAB technique. The authors describe that in both organisms the major component of polar lipids is a methyl ester of PGP (PGP-Me, Fig. 5) instead of the corresponding free acid previously reported.

The glycolipids of halophilic archaebacteria appear to be derived from the basic structure diglycosyl diether, mannosyl-glucosyl-diphytanylglycerol (DGD), by substitution of a sugar or sulfate group at the 6 position of the mannose residue, giving rise to triglycosyl diethers and sulfate diglycosyl (S-DGD-1) diethers, respectively (Fig. 5; Kamekura and Kates 1988).

Triglycosyl glycolipids differ among themselves in the more external sugar residue, which can be galactose (TGD-1) or glucose (TGD-2). TGD-1 in turn originates a sulfate triglycosyl derivative when the sulfate group is linked to C-3 of the third sugar residue (S-TGD-1, Fig. 5). Minor branched tetraglycosyl glycolipid (TeGD) and its sulfated derivative (S-TeGD) show essentially the structure of the sulfate triglycosyl glycolipid with galactose residue (TGD-1), with the addition of α-galactose to the 3 position of the mannose residue (Fig. 5; Kamekura and Kates 1988).

Another sulfated diglycosyl diether, designated as S-DGD-2, has been detected in the extreme halophile strain isolated in Japan, but its structure is yet unknown (Kamekura and Kates 1988).

The membrane lipid composition of nonalkaliphilic halobacteria has proved particularly useful in the taxonomy of such microorganisms. PGS, for example, is not as widely distributed in all strains as PGP and PG. Variations also occur in the type and amounts of the glycolipids; the presence or absence of these compounds and other minor unidentified lipids can be used to rapidly assign an isolate to a particular group. Most *Halobacterium* species are characterized by the presence of PGS, S-TGD-1, and S-TeGD. Members of *Haloarcula* contain PGS as well as TGD-2 as major glycolipids, together with a minor unidentified glycolipid component, and do not possess sulfated glycolipid. The species of the *Haloferax* genus, on the contrary, do not contain PGS but are characterized by the presence of S-DGD-1. The main lipids of the *Halococcus* genus are PG, PGP, TGD-2, and a sulfated diglycosyl lipid, probably S-DGD-1 (Fig. 5; Torreblanca et al. 1986; Kamekura and Kates 1988; Grant and Larsen 1989).

The polar lipid extract from a neutrophilic halophilic archaebacterium isolated from a salt mine contains, in addition to PG, PGP, PGS, S-DGD-1, an analogue of phosphatidic acid (PA, Fig. 5; Lanzotti et al. 1989c), probably unrelated to a biosynthetic route of these complex lipids.

The haloalkaliphilic archaebacteria of the genera *Natronobacterium* and *Natronococcus* have a relative simple polar lipid composition in comparison with

Acidic lipids

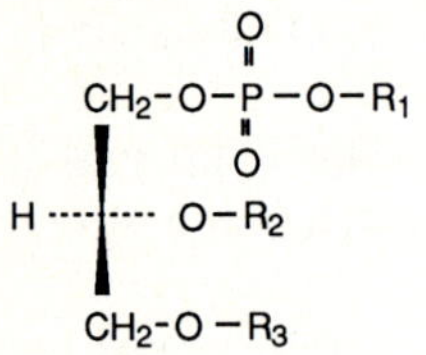

	R_1	R_2	R_3
PG	3'-*sn*-glycerol	C_{20} or C_{25}	C_{20}
PGP	3'-*sn*-glycerol -1'-P	C_{20} or C_{25}	C_{20}
PGS	3'-*sn*-glycerol -1'-sulfate	C_{20}	C_{20}
PA	H	C_{20}	C_{20}
PL2	3'-*sn* -glycerol-1',2'cyclicphosphate	C_{20}	C_{20}
PGP-Me	3'-*sn*-glycerol-1'-P-Me	C_{20}	C_{20}

Glycolipids

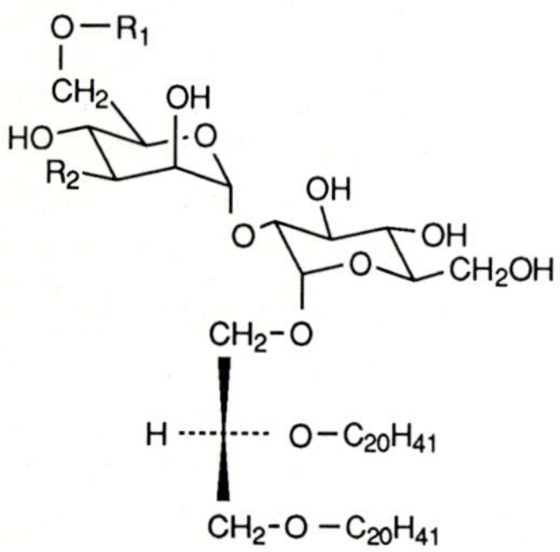

	R_1	R_2
DGD	H	OH
S-DGD-1	$-SO_3H$	OH
TGD-1	ß-gal*p*	OH
TGD-2	ß-glc*p*	OH
S-TGD-1	3-SO_3H-ß-gal*p*	OH
TeGD	ß-gal*p*	O-α-gal*f*
S-TeGD	3-SO_3H-ß-gal*p*	O-α-gal*f*

Fig. 5. Complex lipids in halophiles. *P* Phosphate group

neutrophilic halophiles, in that the glycolipids are completely absent and the major species of polar lipids are generally PGP and PG. However, several minor unidentified phospholipids are also present in examples of these genera. In particular, *Natronococcus occultus* has a significant amount of PGP derivative with a 1′,2′-cyclic phosphate (PL2, Fig 5; De Rosa et al. 1988; Lanzotti et al. 1989a).

In all species of *Natronobacterium* and *Natronococcus*, PGP, the major polar lipid, elutes from the silica column in association with glycine betaine; this last forms an ionic complex with a phospholipid only when two charges are available for complex formation. Thus PG, and PL2, the cyclic derivative of PGP, did not form the complex (De Rosa et al. 1988).

Nicolaus et al. (1989) describe that glycine betaine is membrane-associated; its level increases along with an increase in the PG/PGP ratio when *N. occultus* is grown in different salt concentrations.

At the moment, complex lipids based on $C_{20}C_{25}$ (b, Fig. 2) diether are found only as PG, PGP, and PL2 and are present essentially in haloalkaliphilic species (De Rosa and Gambacorta 1988; Kamekura and Kates 1988).

3.2 Methanogens

Methanogenic archaebacteria have complex lipids based both on diether and tetraether molecules; in this respect they differ from halophiles, which have lipids based only on diethers, and from thermophilic sulfur-dependent species, which have lipids based essentially on tetraethers. However, the knowledge of polar lipids in methanogens is quite incomplete, both because many of these molecules are not fully characterized, and because few species of methanogens are analyzed in detail for their lipid composition.

The complex lipids in methanogens are generally classified by means of specific staining tests into five groups: (1) mean phospholipids; (2) aminophospholipids; (3) aminophosphoglycolipids; (4) phosphoglycolipids; and (5) glycolipids (Nishihara and Koga 1987).

Aminophospholipids were found to be widely distributed in methanogens, in which they prevail in contrast with the other two phenotypes of archaebacteria. For example, the diphytanyl ether analogue of phosphatidylserine (PNL2b, 1, Fig. 6) occurs as a major constituent in the lipids of *Methanobacteriales* (Morii et al. 1986; Koga et al. 1987). In a genus of such an order is also found a phosphoethanolamine derivative of diphytanyl glycerol diether (2, Fig. 6; Kramer et al. 1987; Nishihara et al. 1989).

In species belonging to the *Methanococcus* genus a novel 2,3-di-O-phytanyl-1-(phosphoryl-2-acetoamido-2-deoxy-*β*-D-glucopyranosyl)-*sn*-glycerol has been identified (3, Fig. 4; Ferrante et al. 1986).

Kushwaha et al. (1981) have found in *Methanospirillum hungatei*, a 2,3-di-O-phytanyl-1-(phosphoryl-1′-*sn*-glycerol)-*sn*-glycerol, i.e., the diastereoisomer of PG found in extreme halophiles. In contrast to this result, Ferrante et al. (1987) did not find this phospholipid in *M. hungatei*, but instead, found two new

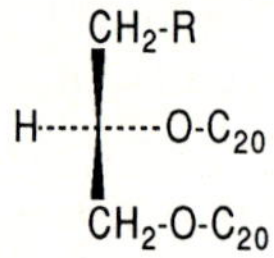

	R		
1	P-CH_2-CH-(NH_3^+)-COOH	7	ß-D-glc*p*-(1-6)-ß-Dglc*p*
2	P-$(CH_2)_2$-NH_3^+	8	α-glc*p*-(1-2)-ß-gal*f*
3	P-1- [2-(NHAc)-2-deoxy] -ß-D-glc*p*	9	ß-gal*f*-(1-6)-ß-gal*f*
4	P-CH- [CH_2-NH^+-$(CH_3)_2$]-$(CHOH)_2$-CH_2OH	10	ß-D-gal*p*-(1-6)-ß-D-gal*p*
5	P-CH- [CH_2-N^+-$(CH_3)_3$]-$(CHOH)_2$-CH_2OH	11	α-D-man*p*-(1-6)-ß-D-gal*p*
6	ß-D-glc*p*	12	P-myoinositol

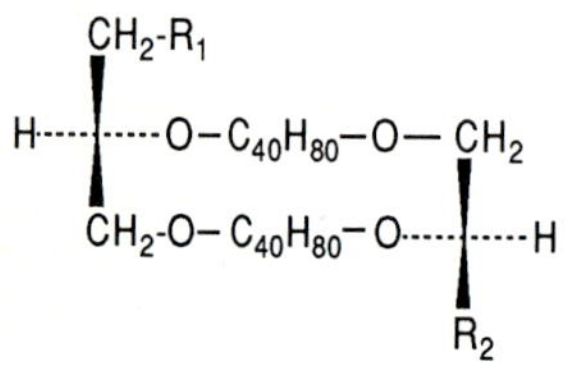

	R_1	R_2
13	OH	CH_2-O-ß-galf-(1-6)-ß-gal*f*
14	OH	CH_2-O-α-glc*p*-(1-2)-ß-gal*f*
15	P-1-*sn*-glycerol	CH_2-O-ß-gal*f*-(1-6)-ß-gal*f*
16	P-1-*sn*-glycerol	CH_2-O-α-glc*p*-(1-2)-ß-gal*f*
17	OH	CH_2-O-ß-D-glc*p*-(1-6)-ß-D-glc*p*
18	P-CH_2-CH-(NH_3^+)-COOH	CH_2OH
19	P-myoinositol	CH_2OH
20	P-$(CH_2)_2$-NH_3^+	CH_2OH
21	P-myoinositol	CH_2-O-ß-D-glc*p*-(1-6)-ß-D-glc*p*
22	P-CH_2-CH(NH_3^+)-COOH	CH_2-O-ß-D-glc*p*-(1-6)-ß-D-glc*p*
23	P-$(CH_2)_2$-NH_3^+	CH_2-O-ß-D-glc*p*-(1-6)-ß-D-glc*p*

Fig. 6. Complex lipids in methanogens. *P* Phosphate group. The core moiety of lipid 10 is depicted in Fig. 2f

aminophospholipids, identified as 2,3-di-O-phytanyl-1-[phosphoryl-2′-(1′-N,N-dimethylamino)-2′,3′,4′,5′-pentane-tetrol]-*sn*-glycerol (PPAD,*4* Fig. 6) and 2,3-di-O-phytanyl-1-[phosphoryl-2′-(1′-N,N,N,trimethylamino)-2′,3′,4′,5′-pentanetetrol]-*sn*-glycerol (PPTAD, *5* Fig. 6).

Glycolipids identified up until now in methanogens are monoglycosyl or diglycosyl derivatives of 2,3-di-O-phytanyl-*sn*-glycerol. The glycosidic residues are glucose for monoglycoside, MGD (6, Fig. 6), and glucose and/or galactose for diglycosides DGD, also named GL1b, DGD-I, DGD-II, (7–9, Fig. 6; Kushwaha et al. 1981; Ferrante et al. 1986; Koga et al. 1987; Nishihara et al. 1989).

In the aceticlastic methanogen *Methanothrix concilii*, Ferrante et al. (1988b) reported the structure of two new glycolipids GL-1, identified as 2-O-phytanyl-3-O-3′-[hydroxy-3′,7′,11′,15′-tetramethyl] hexadecyl-1-O-[β-D-galactopyranosyl-(1–6)-β-D-galactopyranosyl]-*sn*-glycerol (10, Fig. 6) in which the core lipid usually found, 2,3-di-O-phytanyl-*sn*-glycerol, is substituted by the new diether f (Fig. 2); the second glycolipid is based on 2,3-di-O-phytanyl-*sn*-glycerol but the disaccharide moiety is formed by galactose and mannose (11, Fig. 6).

Also identified in methanogens is a phosphatidylinositol derivative of 2,3-di-O-phytanyl-*sn*-glycerol (PL2b or PI, 12, Fig. 6; Koga et al. 1987; Ferrante et al. 1988b; Nishihara et al. 1989).

Kushwaha et al. (1981) have elucidated the structure of complex lipids in *Methanospirillum hungatei*. In this microorganism, the polar complex lipids based on tetraethers are limited to two glycolipids and two phospholipids. In the glycolipids, DGT-I and DGT-II (13,14 Fig. 6), one of the free hydroxyl groups of GDGT a (Fig. 3) is linked glycosidically to a disaccharide residue. From the two glycolipids the phosphoglycolipids PGLI and PGLII (15,16 Fig. 6) are derived in which a 1-*sn*-glycerophosphate residue is attached to the opposite side of the tetraether molecule.

In *Methanobacterium thermoautotrophicum*, in addition to the diether lipids described above (1,2,7,12, Fig. 6), complex lipids based on tetraether a (Fig. 3) are present. The first one, GL1a (17, Fig. 6), is a glycolipid in which one of the free hydroxyl groups of GDGT a (Fig. 3) is linked glycosidically to a glucose disaccharide residue. The others are phospholipids: phosphoserine (PNL2a, 18, Fig. 6), phosphomyoinositol (PL2a, 19, Fig. 6), or phosphoethanolamine (PNL1a, 20, Fig. 6) derivatives of tetraether a (Fig. 3). The latter are phosphoglycolipid derivatives of GL1a with two polar head groups of glucosylglucose and phosphomyoinositol (PGL1, 21, Fig. 6) or phosphoserine (PGL2, 22, Fig.6), or phosphoethanolamine (PNGL1, 23, Fig. 6) separately attached on glycerol moieties (Nishihara et al. 1989).

3.3 Thermophiles

Identities of the polar lipid structures among the thermophilic phenotype are still incomplete, although very often in respect to methanogens, the structures of a whole set of major polar lipids are established.

The complex lipids of thermophilic archaeobacteria are based essentially on tetraethers with the exception of the genera *Thermococcus*, *Pyrococcus*, and an AN1 isolate. These species, which were until now classified in the branch of methanogens and halophiles (Fig. 1; Woese 1987), possess lipids based on 2,3-di-O-phytanyl-*sn*-glycerol (Fig. 2a).

In *Thermococcus* and in *Pyrococcus* a single phospholipid is found (about 85% of total complex lipids), in which the phosphomyoinositol is the polar head attached to the free hydroxyl of the glycerol moiety (12, Fig. 6; De Rosa et al. 1987; Lanzotti et al. 1989b). In addition to this type of phospholipid, which accounts for 40% of total complex lipids, the AN1 isolate has a novel phosphoglycolipid, 2,3-di-O-phytanyl-1-(3-phosphoryl-α-D-glucopyranosyl)-*sn*-glycerol, which represents the first example in archaebacteria of glycolipid with a phosphorylated sugar (Figure not shown; Lanzotti et al. 1989b).

The lipid composition confirms a close phylogenetic relationship among the genera *Thermococcus*, *Pyrococcus*, and the unclassified AN1 isolate, as suggested by molecular analysis (Woese 1987), justifying the introduction of a separate order of *Thermococcales*, which could represent a third major division of the archaebacteria.

In the other thermophilic species, complex lipids based on 2,3-di-O-phytanyl-*sn*-glycerol, are minor compounds which have generally remained unidentified until now.

The complex lipids of *Thermoplasma acidophilum* are based only on GDGTs (Fig. 3) which are differently cyclized. At least six different glycolipids and seven phosphorus-containing lipids have been found, but none of these has been fully characterized. The major component is a phosphoglycolipid derivative, in which a 3-*sn*-glycerol-phosphate and an unidentified monosaccharide are attached to the free hydroxyl groups of the GDGTs (1, Fig. 7; Langworthy 1979). Other minor phosphoglycolipids appear to possess up to three carbohydrate residues along with free amino groups, possibly as amino sugars (Langworthy 1979).

The complex lipids of the thermophiles belonging to the group of sulfur-dependent archaebacteria have now been defined in *Desulfurococcus*, *Thermoproteus*, *Sulfolobus*, *Desulfurolobus*, and *Metallosphaera* (De Rosa and Gambacorta 1988; Thurl and Schafer 1988; Huber et al. 1989). In these microorganisms, the largest percentages of complex lipids occur as phosphoglycolipids, in which a sugar residue and a phosphate group are linked to opposite sides of the tetraether molecules. Polar head groups are restricted to galactose or glucose, or both, and P-myoinositol. Different structures originate from the stereochemistry of the glycosidic bond and the location of the interglycosidic linkage.

In *Thermoproteus* three glycolipids have been found, based on diether a (Fig. 2) and tetraethers a,b,c,f,g (Fig. 3) as core lipids, and glucose residues as polar heads.

The minor glycolipids based on 2,3-di-O-phytanyl-*sn*-glycerol a (Fig. 2) could be probably diether glucosides (Thurl and Schafer 1988).

The tetraether-based glycolipids, named by the authors GL1, GL3, and GL4, have a mono- (2, Fig. 7), di- (3, Fig. 7), and tri- (4, Fig. 7) glucosyl residue as polar head groups, respectively.

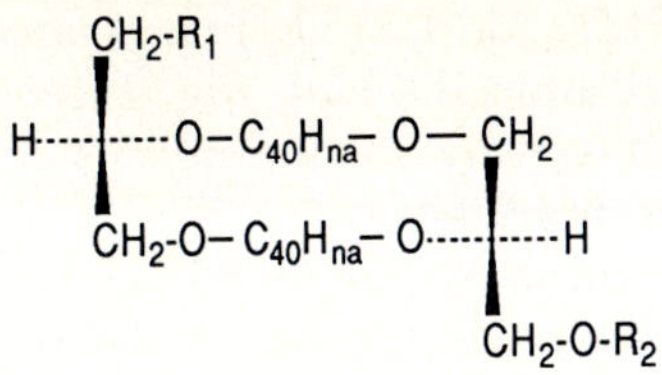

	na	R_1	R_2
1	76-80	P-*sn*-glycerol	monoglycosyl
2	78-80	OH	β-D-glc*p*
3	78-80	OH	β-D-glc*p*-(1-6)-β-D-glc*p*
4	76-80	OH	β-D-glc*p*-(1-6)-β-D-glc*p*-(1-6)-β-D-glc*p*
5	76-80	P-myo-inositol	β-D-glc*p*
6	76-80	P-myo-inositol	β-D-glc*p*-(1-6)-β-D-glc*p*-(1-6)-β-D-glc*p*
7	80	OH	α-glc*p*-(1-4)-β-gal*p*
8	80	P-myo-inositol	β-gal*p*
9	80	P-myo-inositol	α-glc*p*-(1-4)-β-gal*p*
10	72-80	OH	β-gal*p*-β-glc*p*
13	72-80	P-myo-inositol	CH_2OH
14	72-80	P-myo-inositol	β-gal*p*-β-glc*p*

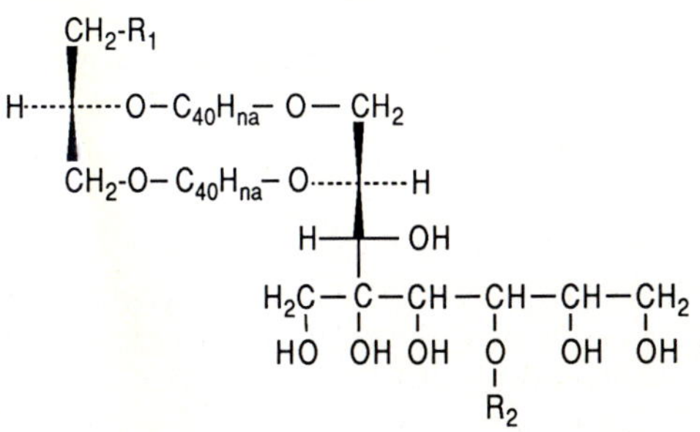

	na	R_1	R_2
11	72-80	OH	β-glc*p*
12	72-80	OH	β-glc*p*-SO_3H-sulfate
15	72-80	P-myo-inositol	β-glc*p*

Fig. 7. Complex lipids in thermophiles. *P* Phosphate group

Phospholipids of this microorganism, named PL3 and PL5 (5,6, Fig. 7), are phosphatidylinositol derivatives of glycolipids GL1 and GL4 (2,4, Fig. 7), respectively. Besides these two phosphoglycolipids, there are small amounts of phospholipids based on both diether and tetraether molecules.

In the *Desulfurococcus* species three complex lipids based on GDGTs have been identified. The first one is a glycolipid, in which one of the free hydroxyl groups of GDGT a (Fig. 3) is linked glycosidically to a glucose-galactose disaccharide (7, Fig. 7). The second is a bipolar phosphoglycolipid, in which one of the free hydroxyls of GDGT is linked glycosidically to galactose and the other one is esterified with phosphomyoinositol residue (8, Fig. 7). The last compound is a bipolar phosphoglycolipid derived from the glycolipid 7 (Fig. 7) in which the free hydroxyl of GDGT moiety is esterified with phosphomyoinositol residue (9, Fig. 7; Lanzotti et al. 1987).

In the *Sulfolobus* species, the complex lipids are based both on GDGTs and GDNTs which are differently cyclized (Fig. 3). The glycolipids are both disaccharide and monosaccharide derivatives of tetraethers. The glycolipid, based on GDGTs, has a glucose-galactose disaccharide glycosidically linked to one of the free hydroxyls of the glycerol moiety (10, Fig. 7). The GDNT-based glycolipid has a glucose-residue attached to the C-6 of the nonitol moiety (11, Fig. 7). From this glycolipid the sulfoglycolipid 12 (Fig. 7) is derived by esterification of the glucose moiety with a sulfate group. The three phospholipids identified in *Sulfolobus* species all contain a myoinositol phosphate residue. The first is derived by esterification of the free hydroxyl group of GDGTs with phosphomyoinositol (13, Fig. 7). The remaining two (14, 15, Fig. 7) are derivative of the glycolipids (10,11, Fig. 7) in which the phosphomyoinositol residue esterifies the free hydroxyl group of the glycerol moiety (Langworthy 1985; De Rosa et al. 1989). In *Sulfolobus* grown autotrophically, two unidentified polar lipids also occur (Langworthy 1985).

In *Desulfurolobus*, grown both aerobically and anaerobically, the same complex lipids are identified as found in the *Sulfolobus* species (10–15, Fig. 7), with the differences in the degree of cyclization of the aliphatic chains (De Rosa and Gambacorta 1988; Trincone et al. 1989b).

In *Metallosphaera*, a new genus ascribable to the order *Sulfolobales*, a lipid pattern of core and complex lipids is found very close to that reported for *Sulfolobus solfataricus* (Huber et al. 1989), although the relative proportions of glycolipids (10,11, Fig. 7) and minor complex lipids are different.

3.4 Complex Lipids and Taxonomy of Archaebacteria

Some general conclusions can be made on complex lipids of the archaebacteria characterized so far. Structural analogies can be seen between complex lipids based on 2,3-di-O-phytanyl-*sn*-glycerol and conventional glycosyl or phosphatidyl diacylglycerols, the glycerol stereochemistry and the presence of ether linkages being the sole differences.

In contrast, tetraether polar lipids are unique in their structure in that the architecture of these bipolar amphipatic molecules has no counterpart in eubacterial and eukaryotic lipids. The complex lipids based on tetraethers may contain polar groups attached to one or both polyol moieties. The largest percentage of these lipids occurs as phosphoglycolipids; there are no examples of symmetrical tetraether molecules.

The relationship between lipid composition and taxonomy in the halophiles has been described above. The polar lipid structures found in halophile phenotypes are unique among archaebacteria. In fact, the phosphoglycerol, as a polar residue of the phospholipids, as well as the mannose as one of the sugar residues of the glycolipids, have never been found in other archaebacteria so far studied; the sole exception is glycolipid 11 (Fig. 6) of the aceticlastic methanogen *Methanothrix concilii* (Ferrante et al. 1988b).

Sulfated complex lipids are also found only in halophiles, with the exception of sulfolipid 12 (Fig. 7) found in *Sulfolobus*, which is a GDNT-based lipid.

Aminolipids are found only in methanogenic archaebacterial phenotypes; they are structurally derived both from diethers and tetraethers. The polar heads in these aminolipids are restricted to ethanolamine and serine, whereas those constituting the glycolipids are limited to glucose and/or galactose. The inositol is a head group of some phospholipids and phosphoglycolipids of the methanogens and is present in all phospholipids of sulfur-dependent archaebacteria. On the basis of structural analyses of a whole set of complex lipids in methanogens, in particular, findings that polar head groups of diether complex lipids are also found in tetraether-based ones, could suggest some direct biosynthetic relationship between diether and tetraether complex lipids (see below; Nishihara et al. 1989).

In thermophilic sulfur-dependent archaebacteria, the glycolipids are present to a lesser amount than acidic lipids. Polar lipids of these organisms are based both on GDGT and GDNT tetraethers. These GDNTs are useful markers for assessing the relatedness of a new organism to the order *Sulfolobales* (Langworthy and Pond 1986; De Rosa and Gambacorta 1988).

4 Biosynthesis of Archaebacterial Ether Lipid

Experiments with labeled acetate or mevalonate have indicated that diether and tetraether lipids of some archaebacteria are derived from acetate via mevalonate by the same general biosynthetic pathway operating in eubacteria and eukaryotes (De Rosa and Gambacorta 1986; Kamekura and Kates 1988). However, in a recent report Ekiel and collaborators (1986) show that, in *Halobacterium cutirubrum* and *Hb. halobium*, methyl and methyne carbons in phytanyl chains are derived from lysine.

The characteristic steps of archaebacterial ether lipid biosynthesis are as follows: (1) ether linkage formation; (2) head-to-head coupling of two geranylgeranyl residues, with reduction to form biphytanyls; (3) cyclization within coupled geranylgeranyl residues, with reduction to form five-membered cyclic biphytanyls; and (4) biogenesis of calditol, the branched chain nonitol occurring in GDNTs (*a′-i′*, Fig. 3) of *Sulfolobaceae*.

4.1 Ether Linkage Formation

The ether linkage formation is the common step in the biosynthesis of all archaebacterial lipids. It was first investigated by Kates and Kushwaha (1978) in *Halobacterium cutirubrum*, using differently labeled glycerols as precursors, and then by De Rosa et al. (1982) in *Sulfolobus solfataricus*.

Kates suggested that a more likely precursor for ether linkage formation would be dihydroxyacetone, which could be formed by the action of a glycerol dehydrogenase, an enzyme known to be active in *Hb. cutirubrum*. Instead, De Rosa et al. (1982) indicated that ether formation takes place without the intervening formation of an oxidized intermediate derivative of the glycerol.

Further biogenetic experiments with 5-^{13}C,5,5-^{2}H mevalonolactone performed on *S. solfataricus* indicate that during the alkylation step of glycerol and nonitol in GDGTs and GDNTs (Fig. 3), both deuterium atoms of the precursor are retained on the first carbons of the isoprenic chains (Trincone et al. 1989a). Poulter et al. (1988) reported data on the incorporation of C_{20} isoprenoidic alcohols with different degrees of unsaturation in *Methanospirillum hungatei*, a strict anaerobic methanogenic archaebacterium. These data indicate that the geranylgeraniol was readily incorporated into diethers and tetraethers of the microorganism, while phytol (with one double bond in the first isoprene unit) and phytanol (without double bonds) were poorly, and not incorporated, respectively.

These results support a prenyl transfer mechanism for the attachment of the isoprene residues to the polyol moiety without double bond isomerization in the first isoprene unit. According to this hypothesis, the unusual configuration of the chiral center in the glycerol moiety of archaebacterial lipids would depend on the stereospecific nature of the alkylation step. In this respect, studies on the chirality of C-2 of the glycerol in ether lipids of archaebacteria have been recently carried out by Kakinuma et al. (1988) with feeding experiments of (RS)-, (R)-, and (S)-(1-$^{2}H_2$) glycerol to the culture of *Hb. halobium*. In these experiments, the *sn*-C-1 of the glycerol moiety of the 2,3-di-O-phytanyl-*sn*-glycerol appears to be derived from the *sn*-C-3 carbon of glycerol.

Given the well-demonstrated ability of prenyl pyrophosphates to act as alkylating agents in other biosynthetic mechanisms, the results support a direct ether bond formation from glycerol (or, facilitated by neighboring group deprotonation, from glycerolphosphate).

4.2 Head-to-Head Coupling

The process of head-to-head coupling is particularly striking and has no parallel in other fields of terpene biochemistry. Up until now there has been no direct information on this mechanism. Incorporation experiments of $^{13}C_2$ acetate (De Rosa and Gambacorta 1986) and 2-^{13}C,2,2-^{2}H mevalonolactone in *Sulfolobus solfataricus* establish that the coupling occurs between the two carbons derived from C-2 of mevalonate and that the metabolic fate of the C-2 hydrogen of the double-labeled mevalonolactone is almost identical, both in the head-to-tail elongation process and in the head-to-head C_{20} coupling (De Rosa et al. 1980a; Trincone et al. 1989a). There is no direct information as to whether the coupling reaction is between C_{20} chains ether-linked to glycerol, or beteween C_{20} precursors themselves. Indirect evidence favors the former: in fact, the structural regularities of GDGTs and GDNTs (Fig. 5) are in accord with the supposition that cyclization, to form cyclopentane rings in C_{40} chains of these tetraethers, occurs in the axially symmetric tetraethers rather than in the free C_{20} or C_{40} components (De Rosa et al. 1980b). Recently, some experimental results on the incorporation of ^{14}C diether in *Methanospirillum hungatei* seem to give an indication that the direct precursor for head-to-head coupling could be an isoprenic C_{20} chain ether linked to glycerol (Poulter et al. 1988).

4.3 Cyclopentane Ring Formation

The high structural specificity of tetraethers (Fig. 3), originating from two polyols and five differently cyclized isoprenoids, with regard to the fact that only nine combinations occur for each tetraether series out of the large number of theoretical possibilities, suggests that some biosynthetic hypotheses for cyclopentane ring formation in *S. solfataricus* would be more plausible than others. For examples if the tetraethers were formed from a pool of mixed C_{40} diols or from variously cyclized diphythanyl polyols, we would expect a much wider range of products. Those actually found all show the same symmetry, with each of the antiparallel C_{40} chains having the same steric relationships to each of the antiparallel polar heads, and cyclizations seem to have been introduced in conformity with that symmetry, probably as a consequence of enzyme-substrate interaction. The regular disposition of cyclopentane rings in these tetraethers could indicate that cyclopentanes were closed in an ordered way by a mechanism that operates in a concerted manner on both alkyl chains, starting from the middle of the isoprenoid system toward ether bonds (De Rosa et al. 1980b).

The mechanism of cyclopentane ring formation (Fig. 8) has been investigated using ^{13}C,^{2}H strategically labeled mevalonolactones as precursors of tetraether lipids in *S. solfataricus*. Results of these experiments allowed the discrimination of a series of mechanisms based on the initial presence of an unsaturation, and possibly concerted with a hydride reduction step (Fig. 8). From the labeling pattern in these experiments, the more plausible mechanism of cyclopentane ring formation is depicted, as indicated in Fig. 8 (Trincone et al. 1989a). However, other mechanisms based, for example, on hydroxylation of isoprenic chains could not be excluded.

4.4 Biogenesis of the Nonitol

GDNT lipids (Fig. 3) have been found only in the *Sulfolobales* and their presence has been a help in ascribing microorganisms belonging to the *Desulfurolobus*,

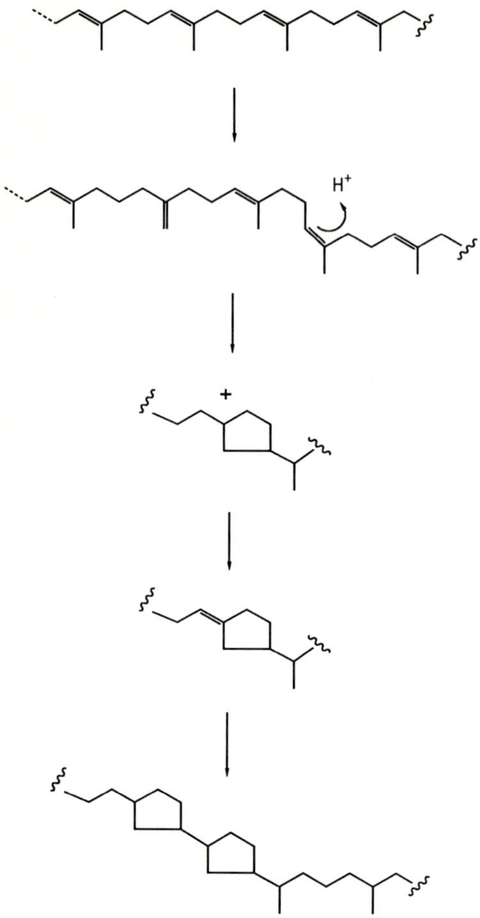

Fig. 8. Biogenetic route for cyclopentane ring formation in the lipids of *Sulfolobus solfataricus*

Acidianus, and *Metallosphaera* genera to this order (De Rosa and Gambacorta 1988; Huber et al. 1989).

The studies of the biogenetic origin and assembly of the carbon skeleton of the nonitol are based on the incorporation in *Sulfolobus solfataricus* of labeled precursors, such as U-^{14}C,2-^{3}H and U-^{14}C, 1(3)-^{3}H glycerols and D- 1-^{14}C,6-^{3}H glucose and fructose. The body of the results, provided by the labeling pattern of the nonitol in GDNTs, leads to the conclusion that, without regarding implications as to stereochemistry or phosphorylation, the biosynthesis of calditol occurs via an aldolic condensation between dihydroxyacetone and fructose, giving rise to a 2-keto-intermediate, which is in turn reduced and alkylated to yield GDNTs (De Rosa and Gambacorta 1986).

4.5 Pathway for Ether Lipid Assembly in Archaebacteria

Biogenetic studies and structural constraints of isoprenoid ether lipids permit the proposition of a general metabolic pattern for the assembly of these molecules (Fig. 9), without considering the implications as the activation of the intermediate.

Ether bond formation between low molecular weight alcohols (glycerol, tetritol, or calditol) and unsaturated isoprenyl pyrophosphate is the common biosynthetic step in the lipid biogenesis of archaebacteria.

In halophiles, the biosynthesis of complex lipids takes place by polar head attachment to the unsaturated intermediates, as indicated by Moldoveanu and Kates (1988).

On the basis of some experiments with labeled phosphorus and some structural regularities of complex lipids of *Methanobacterium thermoautotro-*

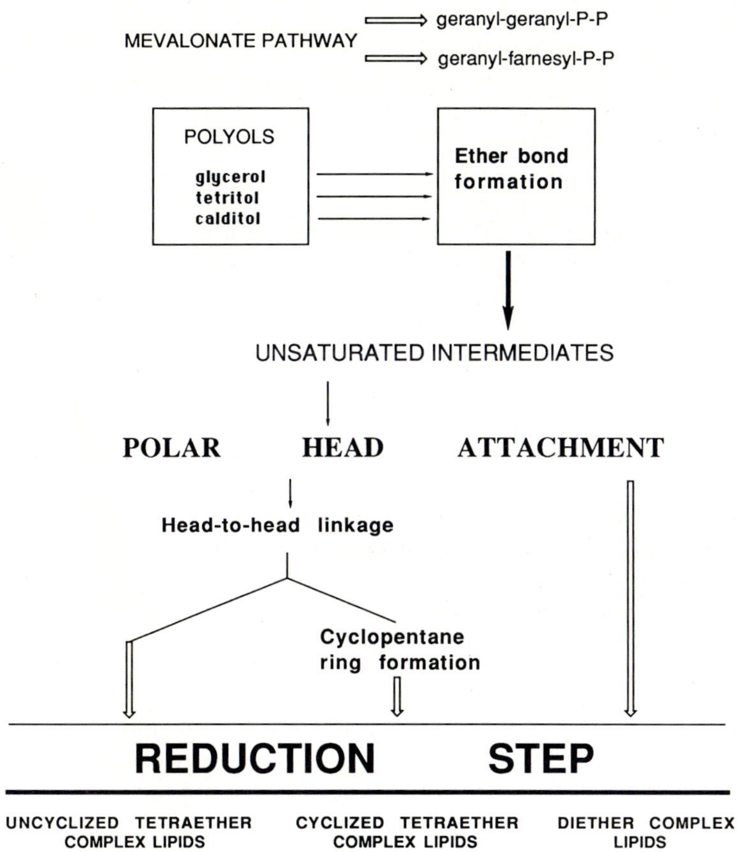

Fig. 9. Pathway for complex lipid biosynthesis in archaebacteria

phicum, Nishihara et al. (1989) propose that biosynthesis of polar tetraether lipids occurs by head-to-head condensation of two molecules of diether complex lipids.

In thermophilic sulfur-dependent archaebacteria, there is no information as to whether the polar head attachment takes place before or after the reduction step.

The question of whether or not a sole general pathway exists for lipid biosynthesis in all archaebacterial phenotypes remains a challenge; further structural and biosynthetic details on complex lipids of archaebacteria are required to answer it.

5 Membranes of Archaebacteria

5.1 Influence of Environmental Stresses on Archaebacterial Membrane Lipids

Many data have been reported on phenotypic changes in the complex lipid composition of eubacteria, but there are few reports of environmental stresses which induce changes in the lipid composition of archaebacteria. In a similar way, much work has been done on the phenotypic adaptation of fatty acyl chain composition in the lipid of eubacteria (Luzzati et al. 1987 and references cited therein), whereas few reports deal with the response of archaebacterial lipid to environmental stresses on the alkyl chains.

In this respect the observation is of interest that when *Sulfolobus solfataricus* and *Thermoplasma acidophilum* are grown at increasing temperatures, the lipids show a higher degree of cyclization of biphytanyl components, which increase with the increasing environmental temperature (De Rosa and Gambacorta 1988). In addition, studies of differential scanning calorimetry of lipids of these microorganisms indicate the presence of a variety of transitions, the critical temperature of which depends on the number of cyclopentane rings, on the isoprenoid C_{40} chains. This evidence suggests that each additional cyclopentane ring in the chain could decrease the available modes of flexing and rotating, and increase the inertial moments of the molecules: the cyclization of the isoprenoid component thus acts as a buffer against the effect of external temperature variation (Gliozzi et al. 1983).

Kushwaha et al. (1982) investigated the effect of salt concentration of growth media on lipids of *H. mediterranei* and *H. cutirubrum*. They showed that among polar lipids the percentage of S-DGD (Fig. 5) increased twice at the expense of PG (Fig. 5) as the total salt concentration was increased from 15 to 30%. In this respect, although the function of sulfated lipids is not clear, some authors speculate that they could play a role in halophiles as a proton donor for the functioning of the purple membrane possessing bacteriorhodopsin (Kamekura and Kates 1988) and/or in the stabilization of the lipid bilayer. In fact, in vitro experiments show that PGP (Fig. 5), one of the major components of lipids of halophilic archaebacteria, gives rise to liposomes with a bilayer structure (Ekiel et al. 1981; Kamekura and Kates 1988) whose stability is strongly affected by the presence of the sulfoglycolipid S-TGD-1 (Fig. 5).

Nicolaus et al. (1989) report that in *Natronococcus occultus*, the relative ratio of PGP/PG increased from 2 to 5 when the salt concentration of the medium increased from 10 to 30% NaCl (w/v) although the total lipid content remained constant. More interestingly, only the levels of the $C_{20}C_{20}$ form of PG and of the $C_{20}C_{25}$ form of PGP are significantly affected by the salt concentration of the medium, resulting in an increase of $C_{20}C_{25}$ alkyl chain-based diether lipids. The results of this experiment support the hypothesis that lipids based on C_{25} chains may have an effect in stabilizing the membranes of haloalkaliphilic archaebacteria.

5.2 Membrane Models in Archaebacteria

With the different structures of the archaebacterial lipids examined so far, the problem arises of how this type of molecule is organized in the membrane.

A series of evidence, such as (1) the structure and dimension of lipids of thermophilic archaebacteria; (2) the absence of a preferential fracture plane upon freeze-fracturing; (3) the extreme rigidity of the thermophilic archaebacterial membranes; (4) labeling experiments with nonpenetrating reagents performed on intact cells of the microorganism; and (5) properties of the black lipid films of GDNTs supports the idea of a monolayer organization of the tetraether lipids in the archaebacterial membranes (De Rosa et al. 1983; Gliozzi et al. 1986).

The results permit the sketching in of hypothetical models of lipid organization in thermophilic, methanogenic, and halophilic archaebacterial membranes, as indicated in Fig. 10. Only in methanogens and in thermophilic archaebacteria does a monolayer membrane occur (Fig. 10c,d). Covalent bonds in fact take place in the middle of all membranes in thermoacidophiles, while a mixed C_{20} and C_{40} ether lipid gives rise to partially covalent bonded monolayer structures in various methanogens, which reflect their lipid composition. The bilayer structure illustrated for extreme halophiles (Fig. 10a), in which, however, the interactions between isoprenoidic components are different from those operating in fatty acyl-based membranes, becomes more strongly connected in haloalkaliphiles, owing to major penetration of the larger C_{25} chains into the opposite lipid layer (Fig. 10b). The increase of C_{25} chain-based lipids increasing the salt concentration of the growth medium in *Natronococcus occultus*, as reported above, supports this hypothesis.

5.3 Survival Strategies at the Level of Archaebacterial Membranes

The uniqueness of these membrane models is remarkable, considering that the organization of the lipids in the fatty acyl-based membranes has so far appeared to be a universally repeated element. The presence of ether lipids in all archaebacteria is an indication that this phenotypic character could give chemical

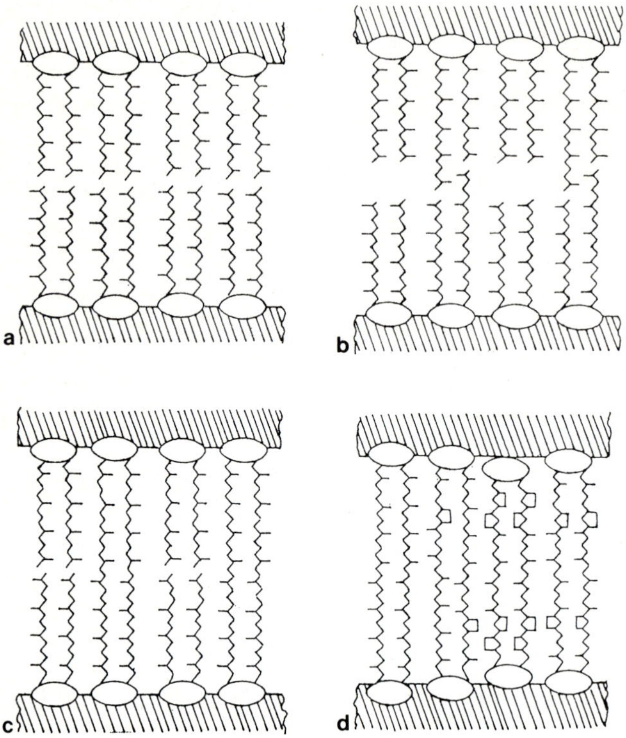

Fig. 10a-d. A sketch of membrane models in archaebacteria. **a** extreme halophiles; **b** alkalophilic halophiles; **c** methanogens; **d** thermoacidophiles

stability to the membrane in extreme environments. However, the comparison of properties of phospholipids containing ester links with their ether-linked analogues indicates that ether linkage has only a small effect on phospholipid packing (Paltauf 1983).

Different strategies have been adopted in different phenotypes of archaebacteria to control the compactness of the membrane, ensuring its optimal functionality in response to environmental stresses. In the alkaliphiles, which must cope with dual stress of high salt concentration and very high pH, an efficient control of the optimal membrane fluidity could be achieved by the extensive use of a C_{25} alkyl chain-based lipid, an alternative to those based on C_{20} alkyl chains.

In a natural environment, in which water is close to the boiling point, a biological membrane based on a lipidic monolayer represents one of the possible strategies to limit lipid mobility, maintaining the fluidity of the system at values compatible with the biological processes. In the classic membranes (Fig. 11), monopolar lipids move into the layer to which they belong. This movement, temperature-dependent, conditions the activity of the membrane in a critical way. In a thermal environment, only the use of a bipolar lipid anchored on two sides

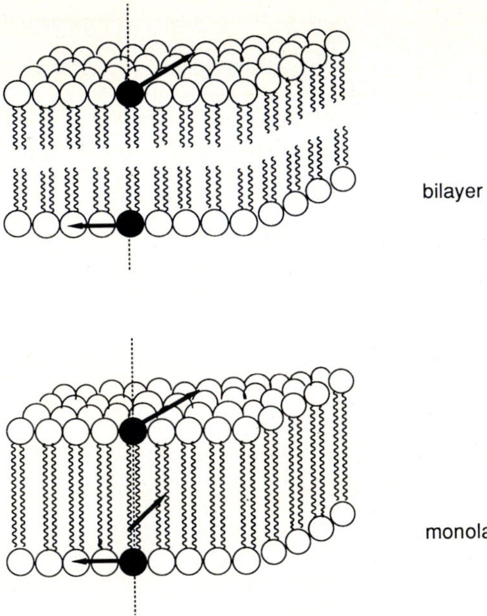

Fig. 11. Degree of freedom of monopolar and bipolar lipids in biological membranes. The *arrows* indicate the movement of lipids along the membrane

of the lipid monolayer permits the control of the lateral creep of the lipids (Fig. 10). Actually, in this manner the mobility of each lipid becomes very limited, because each one of them translates only as much as is permitted by the resultant forces acting on both anchorages, which generally have a low probability of developing in the same direction.

Acknowledgment. The authors thank Mr. Raffaele Turco for artwork.

References

Comita PB, Gagosian RB, Pang H, Costello CE (1984) Structural elucidation of a unique macrocyclic membrane lipid from a new extremely thermophilic, deep-sea hydrothermal vent archaebacterium *Methanococcus jannaschii*. J Biol Chem 259:15234–15241

De Rosa M, Gambacorta A (1986) Lipid biogenesis in archaebacteria. In: Kandler O, Zillig W (eds) Archaebacteria '85. Fischer, Stuttgart, pp 278

De Rosa M, Gambacorta A (1988) The lipids of archaebacteria. Prog Lipid Res 27:153–175

De Rosa M, Gambacorta A, Nicolaus B (1980a) Regularity of isoprenoid biosynthesis in the ether lipids of archaebacteria. Phytochemistry 19:791–793

De Rosa M, Gambacorta A, Nicolaus B, Sodano S, Bu' Lock JD (1980b) Structural regularities in tetraether lipids of *Caldariella* and their biosynthetic and phyletic implications. Phytochemistry 19:833–836

De Rosa M, Gambacorta A, Nicolaus B, Sodano S (1982) Incorporation of labelled glycerols into ether lipid in *Caldariella acidophila*. Phytochemistry 21:595–599

De Rosa M, Gambacorta A, Nicolaus B, Chappe B, Albrecht P (1983) Isoprenoid ethers backbone of complex lipids of the archaebacterium *Sulfolobus solfataricus*. Biochim Biophys Acta 753:249–256

De Rosa M, Gambacorta A, Gliozzi A (1986a) Structure, biosynthesis and physicochemical properties of archaebacterial lipids. Microbiol Rev 50:70–80

De Rosa M, Gambacorta A, Lanzotti V, Trincone A, Harris JE, Grant WD (1986b) A range of ether core lipids from the methanogenic archaebacterium *Methanosarcina barkeri*. Biochim Biophys Acta 875:487–492

De Rosa M, Gambacorta A, Trincone A, Basso A, Zillig W, Holz I (1987) Lipids of *Thermococcus celer*, a sulfur-reducing archaebacterium: structure and biosynthesis. Syst Appl Microbiol 9:1–5

De Rosa M, Gambacorta A, Grant WD, Lanzotti V, Nicolaus B (1988) Polar lipids and glycine betaine from haloalkaliphilic archaebacteria. J Gen Microbiol 134:205–211

De Rosa M, Lanzotti V, Nicolaus B, Trincone A, Gambacorta A (1989) Lipids of archaebacteria: structural and biosynthetic aspects. In: Costa MS, Duarte JC, Williams RAD (eds) The microbiology of extreme environments and its potential for biotechnology. Elsevier Applied Science, London, pp 131

Ekiel I, Mash D, Smallbone BW, Kates M, Smith ICP (1981) The state of the lipids in the purple membrane of *Halobacterium cutirubrum* as seen by ^{31}P NMR. Biochem Biophys Res Commun 100:105–113

Ekiel I, Sprott GD, Smith ICP (1986) Mevalonic acid is partially synthesized from aminoacids in *Halobacterium cutirubrum*: a ^{13}C nuclear magnetic resonance study. J Bacteriol 166:559–564

Ferrante G, Ekiel I, Sprott DJ (1986) Structural characterization of the lipids of *Methanococcus voltae* including a novel N-acetylglucosamine 1-P diether. J Biol Chem 36:17062–17066

Ferrante G, Ekiel I, Sprott JD (1987) Structures of diether lipids of *Methanospirillum hungatei* containing novel head groups N,N-dimethylamino and N,N,N-dimethylaminopentane tetrol. Biochim Biophys Acta 921:281–291

Ferrante G, Ekiel I, Girischandra BP, Sprott DJ (1988a) A novel core lipid isolated from the aceticlastic methanogen *Methanothrix concilii* GP6. Biochem Biophys Acta 963:173–182

Ferrante G, Ekiel I, Girischandra BP, Sprott DJ (1988b) Structure of the major polar lipids isolated from the aceticlastic methanogen, *Methanothrix concilii* GP6. Biochim Biophys Acta 963:162–172

Fredrickson HL, Leeuw JW, Tas AC, van der Greef J, Lavos GF, Boon JJ (1989) Fast atom bombardment (tandem) mass spectrometric analysis of intact polar ether lipids extractable from the extremely halophilic archaebacterium *Halobacterium cutirubrum*. Biomed Mass Spectrom 18:96–105

Gliozzi A, Paoli G, De Rosa M, Gambacorta A (1983) Effect of isoprenoid cyclization on the transition temperature of lipids in thermophilic archaebacteria. Biochim Biophys Acta 735:234–242

Gliozzi A, Bruno S, Basak TK, De Rosa M, Gambacorta A (1986) Organization and dynamics of bipolar lipids from *Sulfolobus solfataricus* in bulk phases and in monolayer membranes. In: Kandler O, Zillig W (eds) Archaebacteria' 85. Fischer, Stuttgart, pp 266

Grant WD, Larsen H (1989) Extremely halophilic archaebacteria. In: Staley JT, Bryant MP, Pfennig N, Holt JG (eds) Bergey's manual of systematic bacteriology, vol 3, Williams and Wilkins, London, pp 2216

Huber G, Spinler C, Gambacorta A, Stetter KO (1989) *Methallosphaera sedula* sp. nov. represent a new genus of aerobic, metal-mobilizing, thermophilic archaebacteria. Syst Appl Microbiol 12:38–47

Kakinuma K, Yamagishi M, Fujimoto Y, Ikekawa N, Oshima T (1988) Stereochemistry of the biosynthesis of *sn*-2,3-O-diphytanyl glycerol, membrane lipid of archaebacterium *Halobacterium halobium*. J Am Chem Soc 110:4861–4863

Kamekura M, Kates M (1988) Lipids of halophilic archaebacteria. In: Rodriguez-Valera F (ed) Halophilic bacteria, vol II. CRC Press, Boca Raton, Florida, pp 25

Kates M, Kushwaha SC (1978) Biochemistry of the lipids of extremely halophilic bacteria. In: Caplan SR, Ginzburg M (eds) Energetics and structure of halophilic microorganisms. Elsevier North Holland Biomedical Press, Amsterdam, pp 461

Koga Y, Ohga M, Nishihara M, Morii H (1987) Distribution of a diphytanyl ether analog of phosphatidylserine and an ethanolamine-containing tetraether lipid methanogenic bacteria. Syst Appl Microbiol 9:176–182

König H (1988) Archaebacteria. In: Rehm HJ (ed) Biotechnology, vol 6. Verlag Chemie, Basel, pp 697

Kramer JKG, Saver FD, Blackwell BA (1987) Structure of the two new aminophospholipids from *Methanobacterium thermoautotrophicum*. Biochem J 245:139–143

Kushwaha SC, Kates M, Sprott JD, Smith ICP (1981) Novel polar lipids from the methanogen *Methanospirillum hungatei* GP1. Biochim Biophys Acta 664:156–173

Kushwaha SC, Juez Perez G, Rodriguez-Valera F, Kates M, Kushner DJ (1982) Survey of lipids of new group of extremely halophilic bacteria from salt ponds in Spain. Can J Microbiol 28:1365–1373

Langworthy TA (1979) Special features of *Thermoplasma*. In: Barile MF, Racin R (eds) The mycoplasma. Academic Press, New York, pp 495

Langworthy TA (1985) Lipids of Archaebacteria. In: Woese C, Wolfe RS (eds) The bacteria. Gustav Fischer, Stuttgart, pp 459

Langworthy TA, Pond JL (1986) Archaebacterial ether lipids and chemotaxonomy. In: Kandler O, Zillig W (eds) Archaebacteria '85. Gustav Fischer, Stuttgart, pp 253

Lanzotti V, De Rosa M, Trincone A, Basso A, Gambacorta A, Zillig W (1987) Complex lipids from *Desulfurococcus mobilis*, a sulfur reducing archaebacterium. Biochim Biophys Acta 922:95–102

Lanzotti V, Nicolaus B, Trincone A, De Rosa M, Grant WD, Gambacorta A (1989a) A complex lipids with a cyclic phosphate from the archaebacterium *Natronococcus occultus*. Biochim Biophys Acta 1001:31–34

Lanzotti V, Trincone A, Nicolaus B, Zillig W, De Rosa M, Gambacorta A (1989b) Complex lipids of *Pyrococcus* and AN1, thermophilic members of archaebacteria belonging to *Thermococcales*. Biochim Biophys Acta 1004:44–48

Lanzotti V, Nicolaus B, Trincone A, De Rosa M, Grant WD, Gambacorta A (1989c) An isopranoid ether analogue of phosphatidic acid from a halophilic archaebacteria. Biochim Biophys Acta 1002:398–400

Luzzati V, Gambacorta A, De Rosa M, Gulik A (1987) Polar lipid of thermophilic prokaryotic organisms chemical and physical structure. In: Engelman DM, Rantez CR, Pollard TD (eds) Annual review of biophysics and biophysical chemistry 16. Annual Review Inc, Palo Alto, CA, p 25

Moldoveanu N, Kates M (1988) Biosynthetic studies of the polar lipids of *Halobacterium cutirubrum* formation of isoprenyl ether intermediates. Biochim Biophys Acta 960:164–182

Morii H, Nishihara M, Ohga M, Koga Y (1986) A diphytanyl ether analog of phosphatidyl serine from methanogenic bacterium, *Methanobrevibacter arboriphilus*. J Lipid Res 27:724–730

Nicolaus B, Lanzotti V, Trincone A, De Rosa M, Grant WD, Gambacorta A (1989) Glycine-betaine and polar lipid composition in halophilic archaebacteria in response to growth in different salt concentration. FEMS Microbiol Lett 59:157–160

Nishihara M, Koga Y (1987) Extraction and composition of polar lipids from the archaebacterium, *Methanobacterium thermoautotrophicum*: effective extraction of tetraether lipids by an acidified solvent. J Biochem 101:997–1009

Nishihara M, Morii H, Koga Y (1989) Heptads of polar ether lipids of an archaebacterium *Methanobacterium thermoautotrophicum*: structure and biosynthetic relationship. Biochemistry 28:95–102

Paltauf F (1983) Ether lipids in biological and model membranes. In: Mangold HK, Paltauf F (eds) Ether lipids: biochemical and biomedical aspects. Academic Press, New York, pp 309

Poulter CD, Aoki T, Daniels L (1988) Biosynthesis of isoprenoid membranes in the methanogenic archaebacterium *Methanospirillum hungatei*. J Am Chem Soc 110:2620–2624

Thurl S, Schafer W (1988) Lipids from the sulfur dependent archaebacterium *Thermoproteus tenax*. Biochim Biophys Acta 961:233–238

Torreblanca M, Rodriguez-Valera F, Juez G, Ventosa A, Kamekura M, Kates M (1986) Classification of non-alkaliphilic halobacteria based on numerical taxonomy and polar lipid composition and description of *Haloarcula* gen. nov. and *Haloferax* gen. nov. Syst Appl Microbiol 8:89–99

Trincone A, Gambacorta A, De Rosa M, Scolastico C, Sydimov A, Potenza D (1989a) Mechanism of cyclopentane ring formation in tetraether lipids of *Sulfolobus solfataricus*. In: Da Costa MS, Duarte JC, Williams RAD (eds) Microbiology of extreme environments and the potential for biotechnology. Elsevier Applied Science, London, pp 180

Trincone A, Lanzotti V, Nicolaus B, Zillig W, De Rosa M, Gambacorta A (1989b) Comparative lipid composition of aerobically and anaerobically grown *Desulfurolobus ambivalens* an autotrophic thermophilic archaebacterium. J Gen Microbiol 135:2751–2757

Tsujimoto K, Yorimitsu S, Takahasi T, Ohashi M (1989) Revised structure of a phospholipid obtained from *Halobacterium halobium*. Chem Commun 668–670

Woese CR (1987) Bacterial evolution. Microbiol Rev 51:221–271

How Nature Engineers Protein (Thermo) Stability

A. FONTANA

1 Introduction

It is generally accepted that the amino acid sequence of a protein determines its unique three-dimensional structure, which in turn dictates the protein biological function (Anfinsen 1973; Anfinsen and Scheraga 1975). At the present, the structures of some 400 globular proteins have been solved by X-ray crystallography; this wealth of structural information has illustrated the subtle ways in which amino acid chains fold into stable globular structures (Richardson 1981). One of the key problems in modern biochemistry and biophysics is to understand the physical principles and forces, as well as mechanistic pathways, leading to folded proteins. This problem is presently the subject of intense research by a great number of investigators using a variety of theoretical and experimental techniques (Creighton 1978, 1985, 1988; Ghélis and Yon 1982; Jaenicke 1987). However, a quantitative understanding is still lacking of the roles of individual amino acid residues in both directing protein folding and stabilizing protein structure. Only the solution of the protein folding and stability problem will pave the way to prediction of the three-dimensional structure of a protein merely on the basis of its known amino acid sequence, as well as to the design of new proteins with desired biological and physicochemical properties (de novo protein design; Salemme 1985; Oxender and Fox 1987; Ohlendorf et al. 1987; De Grado 1988; Goldenberg 1988).

Folded globular proteins are only marginally stable. Their structure is the result of weak noncovalent forces and interactions such as ion pairs, van der Waals contacts, hydrogen bonding, electrostatic forces, hydrophobic interactions, chain entropy, etc. (Tanford 1968, 1970). There is now a consensus that, in the folded protein, the major stabilizing factor is given by the hydrophobic interaction, whereas the major destabilizing effect is the greater conformational entropy of the unfolded polypeptide chain (Baldwin and Eisemberg 1987).

There are several experimental ways to measure protein stability (Schellman and Hawkes 1980). A common procedure would be to expose a protein to a specific environmental condition (heat, acid or alkaline pH, strong denaturant, detergent, etc.) and then to test the recovery of biological activity or of another

Department of Organic Chemistry, Biopolymer Research, Centre of CNR, University of Padua, Padua, Italy.

Guido di Prisco (Ed)
Life Under Extreme Conditions

physicochemical property of the native protein. Unfortunately, quite often the process of protein denaturation is irreversible as a result of an irreversible conformational and/or covalent change (Zale and Klibanov 1983, 1986; Ahern and Klibanov 1985; Klibanov and Ahern 1987; Volkin and Klibanov 1987). In a number of cases, it is possible to (fully) renature the unfolded protein when reexposed to physiological conditions. This reversible denaturation allows an experimental evaluation of the thermodynamic parameters characterizing the unfolding transition, in particular the quantitative measurement of protein stability in terms of ΔG_D, i.e., the difference in the free energy between the folded and unfolded state of the protein in the absence of a denaturant (Pace 1975). A number of quantitative studies of protein stability have been conducted over the years using a variety of globular proteins and employing different physicochemical techniques, the method of choice being differential scanning calorimetry (Privalov 1979, 1982). The striking result emerging from these studies is that the conformational stability (ΔG_D) of globular proteins is remarkably low, usually between 5 and 15 kcal/mol (Pace 1975; Pfeil 1981). It should be stressed that ΔG_D is a measure of the difference of free energy between two protein states (folded and unfolded) and that consequently the stability of a protein may be altered by changing the forces and interactions in both the folded or unfolded form of the protein (Goldenberg 1985).

At present, the problem of protein/enzyme stability (Privalov 1979, 1982; Baldwin and Eisemberg 1987) is gaining widespread interest not only among researchers interested in basic studies of protein structure and folding (Dill 1985, 1987; Schellman 1987), but also among those interested in using enzymes as practical catalysts under different experimental conditions (Torchilin and Martinek 1979; Mozhaev and Martinek 1984; Mozhaev et al. 1988; Shami et al. 1989). In fact, biotechnological applications of enzymes are often hindered by the intrinsic lability of enzymes to environmental factors such as heat, organic solvents, detergents, proteolytic enzymes, etc. The need to overcome these difficulties has been in the past a major incentive in the development of procedures for enzyme immobilization onto polymeric matrices (Chibata 1978) and for stabilizing enzymes by using low molecular weight additives (such as salts, sucrose, amino acids, etc.; Torchilin and Martinek 1979). More recently, genetic engineering techniques have been employed to modify the amino acid sequence of enzymes of industrial interest at specific amino acid residues of the polypeptide chain (site-directed mutagenesis) in order to improve their stability toward heat, pH, and other common protein denaturants (Rastetter 1983; Ulmer 1983). A limitation to the successful design of these new enzyme properties by site-directed mutagenesis is that the effects of amino acid replacements are not easy to predict; thus it is difficult to decide which amino acid substitutions should be made (Fersht et al. 1984; Winter and Fersht 1984; Ackers and Smith 1985; Gerlt 1987). Nevertheless, initial successful studies of protein and enzyme stabilization by using genetic techniques have been already reported in the literature (Perry and Wetzel 1984; Estell et al. 1985; Bryan et al. 1986; Mulkerrin et al. 1986; Wells and Powers 1986; Ahern et al. 1987; Pantoliano et al. 1987a,b; Nicholson et al. 1988; Wetzel 1988; Wetzel et al. 1988; Matsumura et al. 1989a,b; Matsumura et al. 1989b).

An alternative to current efforts in engineering protein/enzyme stability using genetic techniques, and one likely to provide a significant improvement to existing enzyme technology, would be the use of enzymes isolated from thermophilic microorganisms growing optimally at 60–100 °C. Numerous studies carried out in the last two decades on the functional and molecular properties of thermophilic enzymes have clearly established that they are generally much more resistant to heat and most common protein denaturants than their counterparts from mesophilic sources. Thus, considering the biotechnological relevance of enzyme stability, potentialities and advantages of thermophilic (and in general, extremophilic) enzymes for practical applications are becoming more and more clearly recognized, as evidenced by the numerous research groups in both academic and industrial laboratories conducting active research in this field.

There are a number of advantages in using thermophilic enzymes (and the microorganisms from which they are derived; Zeikus 1979) for biotechnological applications (Daniel et al. 1981; Sonnleitner and Fiechter 1983; Fontana 1984; Deming 1986; Wiegel and Ljungdahl 1986). First of all, their enhanced stability and longer half-life than their mesophilic counterparts under different experimental conditions should make enzymatic processes less expensive, by utilizing the mesophilic enzymes presently in use. Bioreactors constituted by immobilized thermophilic enzymes can be operated at a temperature sufficiently high to prevent microbial contamination, which is a major problem in conventional bioreactors operating at room temperature. In general, biochemical processes carried out at high temperature can be more advantageous and efficient, since viscosity is reduced, solubilities are higher, diffusion is accelerated, etc. (Daniel et al. 1981; Hartley and Payton 1983; Sonnleitner and Fiechter 1983). Thermophilic enzymes, which appear to be rather stable to organic solvents (Rella et al. 1987; see also below), will be useful for the bioconversion of lipophilic or water-insoluble compounds, since in this case it is desirable to carry out enzymatic reactions in the presence of organic solvents (Zaks and Klibanov 1985; Deetz and Rozzell 1988). Of note is the fact that potential hazards normally associated in using bacteria are not expected, since thermophiles, and in particular extreme thermophiles, are nonpathogenic, being unable to grow at body temperatures.

But the interest of studies on the stable enzymes and proteins produced by thermophilic microorganisms resides not only in their potential practical applications (Daniel et al. 1981; Sonnleitner and Fiechter 1983; Sonnleitner 1984; Wiegel and Ljungdahl 1986; Rossi 1987, 1988; Lamed et al. 1988). Thermophilic enzymes are powerful, naturally occurring, protein model systems for studying the structural basis of protein stability in general. Indeed, comparative analyses of the molecular structure and properties of the stable thermophilic enzymes (and proteins) and their counterparts from mesophilic sources have been carried out by a number of investigators, allowing one to infer the role of key interactions contributing to protein stability. Thus, the outcome of studies on thermophilic enzymes are strictly related to those of protein engineering studies which employ mutants of proteins and enzymes obtained by genetic methods (Goldenberg 1985, 1988; Leatherbarrow and Fersht 1986; Beasty et al. 1987; Oxender and Fox 1987).

This short chapter summarizes current thinking about molecular mechanisms of protein stability in thermophilic proteins. No attempt is made to provide a complete coverage of the vast literature dealing with thermophilia enzymes, so that the reader may find some personal selection and omissions of issues as well. Many detailed accounts on the properties of thermophilic enzymes can be found in monographs and books (Singleton and Amelunxen 1973; Heinrich 1976; Ljungdahl and Sherod 1976a,b; Zuber 1976, 1981; Brock 1978, 1985, 1986; Friedman 1978; Oshima 1979; Zeikus 1979; Jaenicke 1981; Fontana 1984; 1988; Daniel 1986; Kristjansson 1989). The protein stability problem has been discussed recently in excellent reviews (Tanford 1968, 1970; Pace 1975; Lapanje 1978; Privalov 1979, 1982; Pfeil 1981; Baldwin and Eisemberg 1987; Becktel and Schellman 1987; Schellman 1987).

2 Extreme Environments: the Most Suitable Places to Look for Stable Enzymes

Organisms capable of living at high temperatures have been of great interest to biologists and biochemists, since they exist and are able to grow at temperatures at which their macromolecular constituents, such as proteins and nucleic acids, are expected to become denatured. Over the years, a great number of microorganisms have been isolated and characterized in terms of their optimal temperature of growth (Ljungdahl and Sherod 1976a,b; Brock 1986; Deming 1986). In Table 1, a partial list of the most-studied microorganisms is given together with their maximum (T_{max}) and optimum (T_{opt}) temperature of growth. Thermophilic microorganisms are ubiquitous and have been isolated from a variety of environmental (extreme) conditions, including hot (above 100 °C) and acidic (pH ~ 2) springs, hot water lines, desert sands, etc. Some thermophiles can prosper almost everywhere, as for example the aerobic *Thermus* species which has been also found in natural hot water systems, or the anaerobic *Clostridium thermohydrosulfuricum* which can be isolated from nearly every anaerobic or aerobic soil sample.

There is growing evidence that the upper limit of temperature growth for bacteria is about 110 °C. The exciting report (Baross and Deming 1983) that bacteria isolated from the "black smoker" of the deep sea grow at 250 °C or above at 265 atm has not been confirmed by others (White 1984). In fact, at these extreme temperatures biomolecules (e.g., amino acids) are unstable and are prone to fast degradation, so that life is not possible (Bernhardt et al. 1984; Miller and Bada 1988). A number of interesting thermophiles able to grow at 100 °C or above have been isolated from the fumaroles of the island of Vulcano, Italy (Stetter 1982; Stetter et al. 1987). The barothermophilic archaebacterium *Pyrodictium* species can grow at temperatures up to 105 °C; that seems to be the most extreme thermophile isolated so far (Stetter 1982). With the exception of *Thermotoga maritima* (Huber et al. 1986), all bacteria growing at temperatures about 90 °C are classified as archaebacteria (Woese et al. 1978; Kandler 1981, 1982,

Table 1. Partial list of thermophilic microorganisms and their maximum (T_{max}) and optimum (T_{opt}) growth temperatures

	T_{max}	T_{opt}
Pyrodictium occultum	110	105
Thermoproteus tenax	100	85
Methanothermus fervidus	100	85
Sulfolobus solfataricus	93	87
Thermotoga maritima	90	80
Bacillus caldotenax	85	80
Thermus thermophilus	85	75
Thermus X-1	80	70
Thermus aquaticus	80	70
Caldocellum saccharolyticum	80	68
Clostridium thermohydrosulfuricum	78	69
Bacillus stearothermophilus	78	55–65
Bacillus caldolyticus	76	62
Bacillus thermoproteolyticus	70	50–60
Clostridium thermocellum	68	60
Lactobacillus thermophilus	65	55–63

1984; Woese 1987). These organisms show a plethora of unusual and unexpected characteristics and have been classified as a third primary kingdom, besides eubacteria and eukaryotes (Woese 1987). Among the special characteristics of archaebacteria are the uniquely modified nucleotides, the special structures of cofactors, the unusual cell wall structures and lipids, etc. (Garrett 1985).

Until recently, biochemists used as a source for stable enzymes moderate thermophiles, such as *B. stearothermophilus*, but in recent times much attention is devoted to extreme thermophiles (growing above 70 °C). In general, it has been demonstrated that there is a correlation between growth temperature of a microorganism and thermostability of its enzyme (Oshima 1979; Walker 1979). Thus, in the future, and for the purposes of enzyme technology, we expect much emphasis to be devoted to the cultivation of extreme thermophiles, and in particular archaebacteria, in order to isolate new and generally stable enzymes useful for practical applications.

3 What is Possible to Learn from Structure-Stability Relationships in Thermophilic Proteins?

3.1 Thermophilic Enzymes are Generally Stable

The numerous studies conducted in the last two decades on a variety of enzymes and proteins isolated from many diverse thermophiles have clearly established that these biopolymers invariably show much higher thermal stability with respect to their counterparts from mesophilic sources. Thermophilic enzymes, however, show not only unusual stability to heat, but also to other common protein denaturants such as organic solvents, detergents, chaotropic agents, proteolytic enzymes, extremes of pH, etc. (Veronese et al. 1984). The *general* stability of thermophilic enzymes is extensively documented in the literature, even if systematic studies have not yet been performed to conclude that thermophilic enzymes are always more resistant to all kinds of extreme physical conditions than mesophilic ones. A specific example of the general stability of a thermophilic enzyme is given in Fig. 1, which shows the results of a comparative analysis of the stability of 6-phosphogluconate dehydrogenase from *B. stea-*

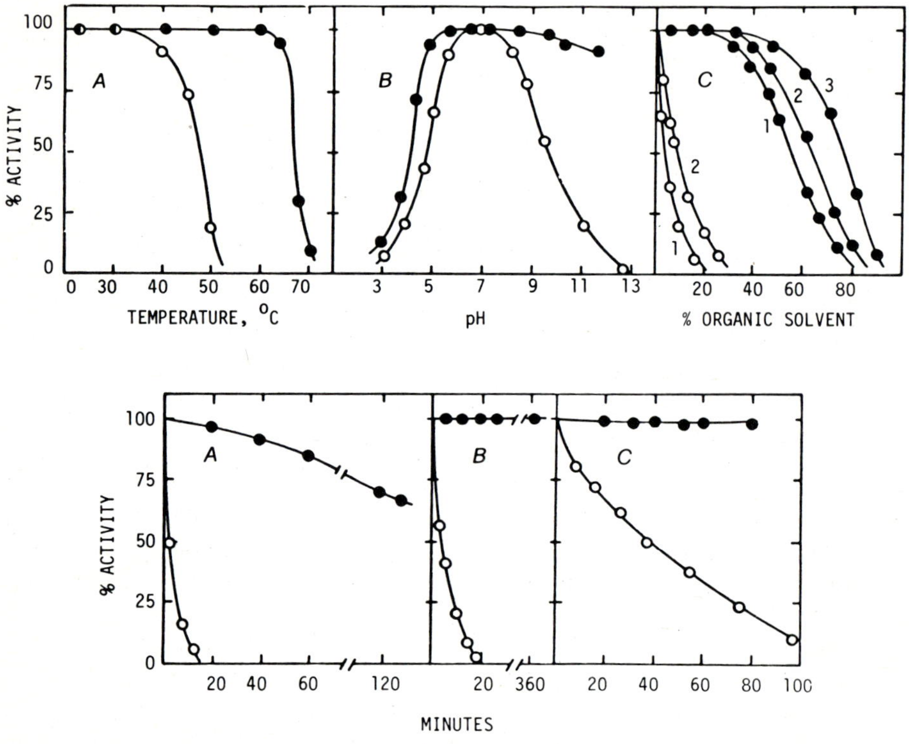

Fig. 1. Effects of experimental conditions on the activity of 6-phosphogluconate dehydrogenase from *Bacillus stearothermophilus* (●) and yeast (○). *Top* The enzyme was dissolved (0.1 mg/ml) in 0.1 M phosphate buffer, pH 7.2, containing EDTA and dithiothreitol (each at 1 mM concentration). **A** The enzyme solution was heated for 15 min at the indicated temperatures and then the residual activity determined at room temperature. **B** The enzyme (0.25 mg/ml) was incubated in universal buffer for 30 min at 20 °C and then assayed for residual activity under standard conditions at pH 8. **C** The effect of organic solvents on the enzyme activity was studied by dissolving the enzyme (0.1 mg/ml) in a mixture of phosphate buffer and organic solvent (*1* acetone; *2* dioxane; *3* dimethylformamide) and measuring the residual activity after incubation in the aqueous organic solvent for 90 min at 20 °C. *Bottom* The enzyme was incubated in phosphate buffer, pH 7.2, at **A** 37 °C plus trypsin, **B** chymotrypsin, and **C** elastase at a protease:enzyme ratio of 1:50 (by weight). The residual activity was measured at the indicated time (After Veronese et al. 1984)

rothermophilus and yeast at a high temperature, in acidic or alkaline pH, in the presence of organic solvents, and toward proteolytic digestion. It is seen that the *B. stearothermophilus* enzyme is invariably more robust than the one from yeast when exposed to the various protein denaturing conditions (Veronese et al. 1984).

3.2 Molecular Mechanism of Protein Stability

The interest of biochemists in thermophilic enzymes has been so far mostly focused on finding molecular mechanisms for their unusual stability (Zuber 1976; Jaenicke 1981; Daniel 1986; Fontana 1988). To this aim, detailed structural studies (amino acid sequence, three-dimensional structure) on these enzymes have been conducted and the results compared with those obtained with mesophilic counterparts. The outcome of these studies was negative, in the sense that contrary to earlier expectations of unusual/special structural features of thermophilic proteins, it was found that thermophilic proteins are not at all unique with respect to their structural properties, but are "typical" proteins characterized by an intrinsic stability, even if in some cases (e.g., α-amylase, thermolysin, caldolysin) protein thermal stability can be achieved extrinsically by the addition of suitable effectors (metal ions, coenzymes, etc.; see below). All current evidence indicates that the enhanced stability of thermophilic enzymes cannot be attributed to a common determinant, but is the result of a variety of stabilizing effects including hydrophobic interactions, ionic and hydrogen bonding, disulfide bonds, metal binding, etc., much the same as had already been observed with mesophilic proteins (Matthews et al. 1974). A quantitative analysis of these factors involved in protein stability appears to be very complicated, since a large number of stabilizing effects must be considered. In addition, since the net free energy of stabilization of a protein (mesophilic and thermophilic) is numerically quite small (5–15 kcal/mol) and the stabilization provided by a single hydrogen bond or salt bridge is of the order of 1–3 kcal/mol, it is clear that interaction energies of this magnitude can arise in different ways through appropriate combinations of the weak bonds commonly involved in stabilizing proteins (Schellman et al. 1981).

Initial efforts in explaining thermostability were carried out by comparing the amino acid composition of two groups of proteins, one from thermophilic and the other from mesophilic organisms (Singleton and Amelunxen 1973; Singleton et al. 1977). The reasoning was that protein thermostability could be explained on the assumption that the necessary and sufficient information for stability of proteins is contained in the amino acid compositions themselves. Hydrophobicity has been thought to play a role in thermostabilization, considering that the strength of hydrophobic interactions increases as a function of temperature, at least up to 60–85 °C (Tanford 1980). Thus, it was anticipated that thermophilic proteins were, in general, more hydrophobic than corresponding mesophilic proteins. A good correlation has been found between certain macroscopic parameters calculated from the amino acid composition of mesophilic versus thermophilic proteins from closely related species (Merkler et al. 1981). These were the hydrophobic index, the ratio of polar to nonpolar residues, the ratio of Arg/(Arg + Lys), plus the percent potential α-helix or β-sheet (α- or β-index). The aliphatic index (the relative volume of a protein occupied by residues with aliphatic side chains: Ala, Val, Ile, Leu) also appears to be related to thermal stability (Ikai 1980). A parameter related to the average residue volume and hydrophobicity also correlates with the transition temperature of 14 nonthermophilic proteins (Bull and Breese 1973). Ponnuswamy et al. (1982) proposed that there groups of residues exist which could either stabilize or destabilize the protein molecule against temperature. The stabilizing group mainly consists of polar-charged and nonpolar residues with aromatic rings/sulfur/nitrogen atoms in their side chains. The polar-uncharged residues mainly destabilize the molecule against temperature, serine being the most destabilizing residue. However, these efforts to relate intrinsic thermostability to global parameters derived only from amino acid composition do not seem to be convincing (Jaenicke 1981; Frömmel and Sander 1989).

The most clear-cut demonstration that *subtle* structural differences between a thermophilic and a mesophilic protein are sufficient to account for their enhanced stability came from the analysis of a series of temperature-sensitive mutants of lysozyme from the bacteriophage T4 (Grütter et al. 1979; Schellman et al. 1981; Alber and Wozniak 1985; Alber et al. 1987; Matthews et al. 1987; Nicholson et al. 1988; Wetzel et al. 1988; Matsumura et al. 1989a,b,c), β-galactosidase from *E. coli* (Langridge 1968), α-subunit of tryptophan synthetase (Yutani et al. 1977, 1984, 1987; Matthews et al. 1980), and others. These studies showed that a substantial effect in the thermal stability of a protein can be achieved simply by a *single* amino acid residue substitution in the protein polypeptide chain. Matthews and co-workers (Grütter et al. 1979) showed that replacement of an Arg by His in T4 lysozyme does not affect the three-dimensional structure of the protein (Remington et al. 1978; Weaver et al. 1989), but lowers its melting temperature by about 14 °C (Hawkes et al. 1989). More recently, many mutants of T4 lysozyme that differ from the wild type in stability have been obtained, and the detailed structural and physicochemical analysis of these mutants have allowed the inference of molecular mechanisms of protein stability (Alber and Wozniak 1985; Alber et al. 1987; Matthews 1987; Karpusas et al. 1989; Matsumura et al. 1989a,b,c).

A detailed examination of sets of homologous proteins from thermophilic and mesophilic sources in terms of amino acid sequences and three-dimensional structures has been a fruitful approach in proposing molecular mechanisms of protein thermostability. For example, it appears that the enhanced stability of thermophilic ferredoxin (Perutz and Raidt 1975) and glyceraldehyde-3-phosphate dehydrogenase (Bieseker et al. 1977; Walker 1979; Walker et al. 1980) in respect to their mesophilic counterparts is due to some extra salt bridges in the thermophilic molecules (Perutz 1978). On the basis of a detailed comparative study of sequences and structures of several thermophilic and mesophilic enzyme molecules, it has been proposed that a decreased flexibility and increased hydrophobicity in α-helical regions are the main stabilizing principles (Argos et al. 1979; Menéndez-Arias and Argos 1989). Frequent amino acid exchanges in thermophilic proteins tend to increase the Ala content of helices. In particular, Ser to Ala and Lys to Arg are the most favorable replacements in thermophilic lactate dehydrogenase, Gly to Ala and Ser to Ala in glyceraldhyde-3-phosphate dehydrogenase, Val to Ala and Lys to Ala in triosephosphate isomerase, etc. (Menéndez-Arias and Argos 1989).

The quite common exchange of Gly with Ala residues in thermophilic enzymes, often located in helical segments (Menéndez-Arias and Argos 1989), has been observed, for example, in neutral protease from *B. stearothermophilus*. It has been attributed to the replacement within a protein α-helix of a poor helix-forming residue (Gly) with a good helix former (Ala; Imanaka et al. 1986). The mutational replacement Gly $\rightarrow$ Ala in an α-helix has been found to increase protein thermostability in λ-repressor (Hecht et al. 1986), as well as to increase the helix propensity in a short peptide such as the C-peptide (13 residues) of ribonuclease (Strehlow and Baldwin 1989). However, an alternative explanation for these observations can be given. The folded form of a protein is destabilized by the effect of the greater configurational entropy of the unfolded protein chain. Thus, a decrease in the magnitude of this entropic contribution is expected to alter ΔG_D of the protein (see above). On this basis, an experimental procedure to enhance protein thermostability by decreasing the entropy of unfolding was proposed by Matthews et al. (1987). Site-directed mutagenesis experiments have been used to prepare a T4 lysozyme mutant with a Gly $\rightarrow$ Ala substitution. This proved to be a stabilizing exchange, as expected from the greater backbone configurational entropy of Gly in respect to Ala, thus requiring more free energy to transfer from the unfolded to the folded state.

The effect in protein stability of a single amino acid replacement may be small and be accounted for a contribution of few kcal/mol to the free energy of folding. However, few of these mutations could be cumulative in their effect and lead to substantial increase in stability (Matthews 1987). This has been proven in a number of cases. For example, an Ala $\rightarrow$ Pro substitution in T4 lysozyme, besides the Gly $\rightarrow$ Ala one (see above), further increases the protein stability; this can be explained in terms of entropic stabilization (Matthews et al. 1987). Two individual mutations that increase thermostability in kanamycin nucleotidyltransferase have been shown to have a cumulative effect when they are concurrently introduced into the protein (Matsumura et al. 1986). The cumulative effect of residue mutations can also explain the fact that a single destabilizing point mutation in neutral protease from *B. stearothermophilus* can be compensated by two stabilizing point mutations (Imanaka et al. 1986). Six individual amino acid substitutions at separate positions in the tertiary structure of subtilisin BPN′ were found to increase the stability of the enzyme (Pantoliano et al. 1989). Individually, the six amino acid substitutions increase ΔG of unfolding between 0.3 and 1.3 kcal/mol, and the combination of them give a net increase of 3.8 kcal/mol. The mutant subtilisin with six residues replaced is ~300 times more resistant to inactivation. More recently, the introduction into the T4 lysozyme molecule of a number of disulfide bonds was shown to dramatically increase the protein thermal stability (Matsumura et al. 1989b). A triple-disulfide variant of T4 lysozyme unfolds at a temperature 23.4 °C higher than the wild-type lysozyme.

3.3 Ion Binding Affects Protein (Thermo)Stability

Ligand binding is a general phenomenon of protein stabilization observed in both mesophilic and thermophilic proteins (von Hippel and Wong 1965; Arakawa and Timasheff 1982). Among the many examples of stabilization of proteins by bound ions, we may recall the examples of parvalbumin, α-lactalbumin, calmodulin, troponin, superoxide dismutase, alkaline phosphatase, and others (see Pantoliano et al. 1988, for some references). Often proteins are stabilized against thermal inactivation when heated in the presence of relatively high salt concentrations. Just to mention an example from the recent literature, D-glyceraldehyde-3-phosphate dehydrogenase from both *Thermoproteus tenax* (Hensel et al. 1987) as well as *Methanococcus fervidus* (Fabry and Hensel 1987) is strongly stabilized against thermal inactivation in the presence of 0.1–2 M salts.

The role of calcium in protein stability has been documented in a number of proteins and enzymes (Ginsburg and Carrol 1965; von Hippel and Schleich 1969). One of the most well-studied cases is the family of homologous neutral proteases produced by mesophilic (*B. cereus*, *B. subtilis*, *B. amyloliquefaciens*) and thermophilic (*B. thermoproteolyticus*, *B. stearothermophilus*) bacteria. These proteases share a number of functional properties in common, as expected from the fact that their amino acid sequences show extensive homologies (Kubo and Imanaka 1988, and references cited therein). These similarities include optimum pH of activity, substrate specificity, molecular weight, absence of thiol or disulfide groups, and binding of zinc and calcium ions. It was found that a zinc ion is located at the active site of the enzyme, whereas two to four calcium ions are bound at different locations in the folded enzyme and seem to play a prominent role in protein stability (Roche and Voordouw 1978). Thermolysin, the neutral metalloendopeptidase from *B. thermoproteolyticus*, has been the subject in recent years of intensive research, which has allowed the analysis of its amino acid sequence (Titani et al. 1972), three-dimensional structure (Colman et al. 1972; Matthews et al. 1972, 1974; Holmes and Matthews 1982), mode of binding of inhibitors (Tronrud et al. 1986), and, in particular, the stereochemistry of the protein-bound single zinc and four calcium ions (Roche and Voordouw 1978). Thus, thermolysin can be considered the best-characterized thermophilic protein and thus a most suitable protein model to address problems of structure-stability relationships in proteins. In fact, on the basis of the features of the X-ray structure of thermolysin, Matthews et al. (1972, 1974; Colman et al. 1972) were the first to propose that thermophilic proteins do not need to be unusual in their structural characteristics with respect to mesophilic counterparts, and that protein thermal stability should arise from a proper contribution of the different weak forces and interactions which stabilize proteins in general. The subtle structural differences between thermophilic and mesophilic proteins were also apparent after the comparative analysis of the X-ray structure of thermolysin and the mesophilic neutral protease from *B. cereus* (Pauptit et al. 1988). This analysis allowed an evaluation of current views and proposals on the mechanisms of protein thermostability. Of note, the less stable *B. cereus* protease, among the several residue exchanges with respect to thermolysin, shows two peculiar substitutions of Gly residues in place of Ala residues located in positions 141 and 287 of the polypeptide chain, both comprising helical segments in both proteins. Again, we can observe that the stabilizing Gly → Ala substitution has been used by nature to engineer protein thermostability in neutral proteases (see above).

The role of calcium in stabilizing the thermolysin molecule was the subject of intensive investigations (Drucker and Borchers 1971; Feder et al. 1971; Voordouw and Roche 1975; Dahlquist et al. 1976; Fontana et al. 1976, 1977; Tajima et al. 1976; Voordouw et al. 1976). The stabilizing effect of calcium on thermolysin was apparent from direct experiments, as well as inferred from the fact that thermolysin binds four calcium ions per molecule, whereas the less stable mesophilic protease binds only two ions. The role of the individual calcium ions in thermolysin was also investigated employing physicochemical and protein chemistry methods. One of the four calcium ions is easily removed from thermolysin by soaking crystals of the protein in the presence of mM ethylenediaminotetraacetic acid (EDTA), as given by difference crystallographic maps (Weaver et al. 1976). These crystallographic data allowed Matthews and coworkers to conclude that the calcium ion (Ca-4) bound at Asp^{200} is the most easily displaced by the chelating agent. When an analogous experiment was performed in solution, it was found that treatment of thermolysin in the presence of 10 mM EDTA leads to rapid enzyme inactivation as a result of an almost quantitative autolysis at peptide bonds Gly^{196}-Ile^{197} and Ser^{204}-Met^{205}. This leads to a *nicked* thermolysin species constituted by three fragments (1–197, 198–204, and 205–316) associated in a stable folded complex (Fassina et al. 1986). Taking together the effects of EDTA on thermolysin in the crystal and in solution, it was inferred that the most easily (with mM EDTA) displaced calcium ion is Ca-4, and that this ion, bound to the protein, protects the corresponding ion binding loop from autolytic inactivation (Weaver et al. 1976; Fassina et al. 1986; Fontana et al. 1986b). This is at variance from an earlier proposal that the calcium ions at the "double site" dissociate more easily from the protein (Voordouw and Roche 1975).

A recent study on the role of calcium on the stability of thermitase, an extracellular protease isolated from *Thermoactinomyces vulgaris*, was recently reported (Frömmel and Sander 1989). Thermitase is the most thermostable member of the subtilisin family, which includes also subtilisin Carlsberg, subtilisin Novo, and proteinase K. Since the amino acid sequences of these homologous proteins are known, as well as the three-dimensional structures of all but thermitase, this set of proteases is also an excellent model to study structure-stability properties, much the same as the neutral protease family (see above). Moreover, since subtilisin shows considerable practical interest (the world production is of several hundred tons per year; Hartley 1986) and has been the subject of detailed protein engineering studies (Estell et al. 1985; Bryan et al. 1986; Pantoliano et al. 1987a,b), it is of interest to uncover the molecular properties of thermitase which dictate its enhanced stability. Frömmel and Sander (1989) first predicted a three-dimensional model of thermitase based on sequence homology and then confronted some proposals for protein thermostability with the molecular details of the protease. The result of this analysis allowed the authors to exclude that the stability of thermitase arises from salt bridges and hydrophobic interactions, proposing that an unusually tight binding of calcium is most likely the major cause for the thermostability of thermitase (Frömmel and Höhne 1981). Of interest is that the location of a calcium ion bound near Asp^{57} to the protein molecule was predicted by the authors, and was subsequently confirmed by X-ray analysis.

3.4 Enzymes Can Be Cold Inactivated

To document how subtle the balance is of stabilizing and destabilizing interactions in globular proteins, it is worth mentioning here that proteins can be inactivated also by *cold*. The low-temperature protein unfolding/inactivation has been shown to occur in a number of cases and is assumed to derive from the fact that proteins show positive values for ΔC_p, which is the difference in heat capacity between the native and unfolded forms and is considered to arise from hydrophobic interactions (Privalov 1979; Tanford 1980; Baldwin 1986; Privalov et al.1986; Schellman 1987; Griko et al. 1988; Pace and Laurents 1989). Several mesophilic proteins were shown to be inactivated by cold (e.g., β-lactoglobulin, α-chymotrypsinogen, ribonuclease, metmyoglobin, T4 lysozyme; see Chen and Schellman 1989, for references), as well as phosphoglycerokinase from *Thermus thermophilus* (Nojima et al. 1977).

3.5 Thermophilic Enzymes Are Rigid Molecules

Thermostable proteins from thermophiles are molecules quite rigid at room temperature with respect to their mesophilic counterparts, a direct consequence of the "clamping" effect of protein structure brought about by the interactions and forces which stabilize thermophilic proteins. This rigidity has an adverse effect on the catalytic efficiency of thermophilic enzymes, since it is known that an appropriate degree of flexibility is required for enzyme catalysis (Zuber 1981). In fact, considering that the rates of enzyme reactions normally double on being performed at a temperature 10 °C higher, one would expect that thermophilic enzymes should have an extremely high activity at thermophilic temperatures. In contrast, it has been observed that the specific activity of thermophilic enzymes is less than would be predicted from the specific activity of their mesophilic counterparts. Only at high temperature are thermophilic enzymes sufficiently flexible to be fully active and yet rigid enough not to be denatured. This low catalytic efficiency of thermophilic enzymes at room temperature has been quite often observed (Zuber 1981; Brock 1985).

Enzyme catalysis is a complex process, involving specific binding of the substrate, participation of functional (acid or basic) groups in catalysis characterized by specific pK values, a microenvironment at the active site of defined dielectric constant, etc. Thus, the molecular mechanism of enzyme catalysis is the result of a subtle balance between a variety of energetic and structural effects, so that perhaps it is not appropriate to overemphasize the role of protein motility in catalysis. Nevertheless, we may recall the *activation* of thermophilic enzymes which are quite often observed in the presence of protein denaturants (salts, organic solvents, guanidine hydrochloride) at moderate concentrations. For example, glyceraldehyde-3-phosphate dehydrogenase from *Thermus thermophilus* is activated when salts and ethanol are added to the enzymatic assay solution (Fujita et al. 1976) and 6-phosphogluconate dehydrogenase from *B. stearothermophilus* is more active in the presence of organic solvents (dioxane, dimethylformamide, acetone; Veronese et al. 1984). A number of organic solvents have been shown to substantially activate the malic enzyme from the extreme thermoacidophilic archaebacterium *Sulfolobus solfataricus* (Guagliardi et al. 1989). It was shown that β-galactosidase from *Sulfolobus solfataricus* is ~30% activated by heating at 75 °C, and two- to threefold activated when assayed in the presence of 0.05–0.15% sodium dodecyl sulfate; this activation appears to result from conformational changes of the enzyme, as detected by circular dichroism and fluorescence measurements (Rella et al. 1988). The catalytic activity of malate dehydrogenase isolated from moderately thermophilic, as well as extremely thermophilic, bacteria is strongly (up to 15 times) activated, if the assays are performed in the presence of 0.4–0.8 M KCl, 3–20% acetone, 4–8 M urea, 0.5–2 M guanidine hydrochloride (Sundaram et al. 1980). These various observations can be taken as an indication that thermophilic enzymes are activated by some loosening of their structure in the presence of protein perturbants. Moreover, the strict inverse correlation between enzyme activity and thermostability has been demonstrated in few cases. The specific activity (i.e., catalytic efficiency) of enolase from rabbit, yeast, *Thermus* X-1 and *Thermus aquaticus* YT-1 at room temperature is of the inverse order of their thermostability (Barnes and Stellwagen 1973; Stellwagen et al. 1973). Similarly, this inverse correlation was also elegantly demonstrated by measuring stability-activity relationships in four mutants of kanamycin nucleotidyltransferase with single or double amino acid replacement(s) (Matsumura et al. 1986). On the other hand, the stabilization of T4 lysozyme by the introduction of multiple disulfide bonds does not impair the enzyme activity (Matsumura et al. 1989c). Similarly, a thermostable variant of subtilisin shows both enhanced thermal stability and catalytic potency (Bryan et al. 1986).

The rigidity of thermophilic proteins has been verified by utilizing hydrogen exchange measurements (Tsuboi et al. 1978) and theoretical calculations (Vihinen 1987). The H-D exchange rates in globular proteins are a reflection of protein motility/flexibility, as documented, for example, by the inverse correlation between the melting temperatures (T_m) and H-D exchange rates in a series of derivatives of basic pancreatic trypsin inhibitor (Wütrich et al. 1980). The higher the exchange rate, and thus the higher the flexibility, the lower is the denaturation temperature. Elongation factor Tu of *Thermus thermophilus* shows a reduced rate of exchange with respect to the more thermolabile *E. coli* factor (Tsuboi et al. 1978). The increased rigidity of thermophilic proteins with respect to their mesophilic counterparts was also documented by calculating protein flexibility indices (F) derived from normalized B-values (temperature factors determined crystallographically) of individual amino acids in several refined three-dimensional structures of globular proteins. The results of these calculations showed that F-values correlate with protein stability, i.e., that rigidity correlates with thermostability of proteins (Vihinen 1987).

That the stable proteins from thermophilic microorganisms are rigid molecules was also evidenced while using proteolytic enzymes as probes of protein motility (Fontana 1989; Fontana et al. 1989). In analogy to all enzymatic reactions, the proteolytic cleavage of a polypeptide chain occurs only if the site of cleavage can bind and adapt itself in a specific way to the stereochemistry of the active site of the protease. This is difficult to achieve with native globular proteins, whereas denatured/unfolded proteins are much more susceptible to proteolysis. On this basis, it was proposed that only the unfolded species of globular proteins are attacked by proteases, and that the actual equilibrium between the native (N) and denatured (D) form of a globular protein controls the rate of protein degradation. The N-D equilibrium is expected to be shifted towards the N form under physiological conditions with the stable thermophilic proteins, so that proteolysis would be hampered. Indeed, exceptional resistance to proteolysis of thermophilic enzymes has been documented in a number of cases. As shown in Fig. 1, 6-phosphogluconate dehydrogenase from *B. stearothermophilus* is much more resistant at room temperature to proteolytic action in respect to the counterpart from yeast (Veronese et al. 1984). Similarly, the activity loss due to proteolysis of asparaginase and β-galactosidase from thermophilic and mesophilic sources correlates with the thermal instability of the enzymes (Daniel et al. 1982). Oligo-1,6-glucosidase from *B. thermoglucosidius* was more resistant against proteolysis than its homologous counterpart from *B. cereus* (Suzuki and Imai 1982).

A systematic study of the behavior of thermophilic enzymes towards proteolysis was conducted, using the metalloendopeptidase thermolysin as a model protein (Vita et al. 1985; Fontana 1989; Fontana et al. 1986a,b, 1989; Fassina et al. 1986). In this case, specific experimental conditions were found which allowed limited proteolysis by added protease or autolysis of thermolysin. It was possible to isolate and characterize several "nicked" thermolysin species which were constituted by two as well as three fragments, associated in stable complexes (Fontana et al. 1986a, 1989). The pattern of protein fragmentation of the thermolysin molecule observed under different experimental conditions permitted the inference that some molecular aspects of the protein-protein recognition process underlie the proteolytic event and, in particular, established that exposed and flexible loops of the thermolysin molecule are the most susceptible sites of proteolysis. Of note, a striking correlation between sites of proteolysis and sites characterized by high mobility in the thermolysin chain was observed, these last determined crystallographically (B-values; Fontana et al. 1986a).

The correlation between protein stability and resistance to proteolysis is substantiated by the fact that proteins with relatively short half-lives in vivo have high protease susceptibility and are generally unstable in vitro (Goldberg and Dice 1974). An interesting study verifying the correlation between protein stability (and thus rigidity) and proteolytic degradation was reported by Yutani and co-workers (Ogasahara et al. 1985). The rate of proteolytic degradation by trypsin, subtilisin, and pronase of mutants of the α-subunit of tryptophan synthase from *E. coli* was higher for the less stable mutants. Similar observations were reported for mutants of kanamycin nucleotidyltransferase (Matsumura et al. 1986) and T4 lysozymes (Schellman 1986).

3.6 Putting Thermophilic Genes into Mesophiles and Vice Versa

Genetic engineering techniques currently employed to facilitate production and usefulness of mesophilic enzymes are also applicable to thermophilic enzymes. In fact, it has been amply demonstrated that genes from thermophiles can be expressed in mesophiles such as *E. coli*, resulting in the production of thermostable enzymes, thus again demonstrating an intrinsic stability of thermophilic enzymes (Nagahari et al. 1980; Tanaka et al. 1981; Tsukagoshi et al. 1984; Joliff et al. 1986; Schwarz et al. 1986; Sekiguchi et al. 1986a,b; Love and Streiff 1987; Fabry et al. 1988; Nagata et al. 1988). This opens the way to the production of the desired stable enzymes, using well-known mesophilic microorganisms and cultivation techniques. It is also of interest to observe that, since the cellular proteins of *E. coli* are heat-denatured and precipitated, heat treatment of *E. coli* cell extract containing the thermostable enzyme can provide a simple system for a partial and easy purification of the desired enzyme (Tanaka et al. 1981).

Thermophilic microorganisms could be very important vehicles for the production of stable enzymes using genetic engineering methods. In fact, the method has been described for rapidly generating thermostable enzyme variants, introducing the gene coding for a given mesophilic enzyme into a thermophile (*B. stearothermophilus*), and then selecting variants which retain the enzymatic activity at the higher growth temperature of the thermophile (Matsumura and Aiba 1985; Liao et al. 1986). The development of new host-vector systems in thermophilic bacteria has been achieved; few structural genes for extracellular enzymes from both a mesophile and a thermophile have been cloned and expressed in *B. stearothermophilus* at elevated temperatures (Imanaka 1983). Using these procedures, thermostable variants of kanamycin nucleotidyl transferase have been successfully produced, involving single amino acid replacement with respect to the wild-type enzyme (Matsumura and Aiba 1985). These results clearly demonstrate the possibility of introducing foreign DNA into *B. stearothermophilus*, thus extending the range of enzymes whose thermostability can be increased by selection in this thermophile.

4 Outlook

The reasons for the continuing interest in thermophilic enzymes is related not only to the scientific curiosity of understanding the molecular mechanism by which these macromolecules maintain their structure and function at high temperature, but also to the practical applications and developments connected with enzyme technology (Daniel et al. 1981; Sonnleitner and Fiechter 1983; Wiegel and Ljungdahl 1986; Saha and Zeikus 1989; Oshima and Soda 1989; Lamed et al. 1988; Kristjanson 1989). The exploitation of some stable enzymes isolated from thermophilic microorganisms is already documented (proteases, dehydrogenases, amylases, glucose isomerases), but the discussion of this aspect is beyond the scope of this report. Here it suffices to recall that the recent development of the "polymerase chain reaction" (PCR) technique for the amplification of DNA fragments was made possible by the availability of the highly thermostable DNA-polymerase from *Thermus aquaticus*, since the "thermal cycle" involves heat denaturation of DNA at high temperature (Saiki et al. 1988). The PCR technique allows molecular biologists to produce microgram quantities of DNA within a few hours (instead of days-weeks) from just picogram amounts of starting material. The technique, which is now automated, represents a cell-free cloning system and thus a true revolution in molecular biology, with extremely important research and clinical applications (direct sequencing of genomic DNA, prenatal diagnosis, detection of viral pathogens, analysis in forensic medicine, etc.; Oste 1988).

Since the metabolic activities of thermophilic bacteria are similar to that of the mesophilic ones, it is expected that any enzyme already found in a mesophilic bacterium will most likely also be found in a thermophilic one. Thus, considering the wide variety of bacteria living in extreme environments at about 100 °C or above (Zeikus 1979; Brock 1986), it appears that there is an unlimited variety of stable enzymes that can be isolated and successfully exploited in biotechnology (Kristjanson 1989). Besides their possible industrial applications, thermophilic enzymes have a number of useful applications also for basic research, since they are the most suitable protein models for addressing a number of relevant problems of contemporary interest in protein stability and folding. Moreover, they are good subjects for experimentation under conditions in which other mesophilic proteins denature; thus they allow investigations over a wide range of temperatures and other physical conditions.

Genetic studies on thermophilic microorganisms (Imanaka 1983; Matsumura and Aiba 1985; Liao et al. 1986; Oshima and Soda 1989) are being conducted in a number of laboratories. There are reasons to believe that this area of research will expand dramatically in the very near future. In fact, it has been amply demonstrated that thermophilic genes can be introduced into mesophiles in order to produce and isolate stable enzymes of practical interest (pullulanase, β-glycosidase, leucine dehydrogenase, esterase, glucose isomerase, neutral protease, alanine racemase, leucine dehydrogenase, amylase, etc.) (Tsukagoshi et al. 1984; Hirata et al. 1985; Iijima et al. 1986; Joliff et al. 1986; Schwarz et al. 1986; Love and Streiff 1987; Fukusumi et al. 1988; Fabry et al. 1988; Nagata et al. 1988; Oshima and Soda 1989). The genetic engineering techniques that are in current use to enhance the usefulness of the less stable enzymes from mesophiles can also be applied to those of the (extreme) thermophiles, thereby extending their range of possible applications (Gerlt 1987).

The outcome of additional studies aimed at a deeper understanding of the structure-stability relationships in (thermo)stable proteins will allow the identification of possible strategies for improving the stability of proteins in general by using genetic techniques such as those of site-directed mutagenesis (Winter and Fersht 1984; Leatherbarrow and Fersht 1986; Gerlt 1987; Matthews 1987; Shaw 1987; Goldenberg 1988; Nosoh and Sekiguchi 1990). In this respect, we may observe that presently a "flood" of new proteins related to wild-type species are being produced by genetic methods in many laboratories (Oxender and Fox 1987). Perhaps we could consider that there is no need to further extend the large number of protein mutants already available. Instead, more intensive and detailed studies, using a variety of theoretical and experimental approaches, are required to extract the full information present in both the available protein mutants produced by genetic methods, as well as the great variety of stable proteins produced by thermophilic microorganisms. Thus, by studying the properties of thermophilic proteins we will learn how efficient protein engineering has occurred in nature, since life began by random mutation and natural selection of protein molecules.

References

Ackers GK, Smith FR (1985) Effects of site-specific amino acid modification on protein interactions and biological function. Annu Rev Biochem 54:597–629

Ahern TJ, Klibanov AM (1985) The mechanism of irreversible enzyme inactivation at 100 °C. Science 228:1280–1284

Ahern TJ, Casal JI, Petsko GA, Klibanov AM (1987) Control of oligomeric enzyme thermostability by protein engineering. Proc Natl Acad Sci USA 84:675–679

Alber T, Wozniak JA (1985) A genetic screen for mutations that increase the thermal stability of phage T4 lysozyme. Proc Natl Acad Sci USA 82:747–750

Alber T, Dao-pin S, Nye JA, Muchmore DC, Matthews BW (1987) Temperature-sensitive mutations of bacteriophage T4 lysozyme occur at sites with low mobility and low solvent accessibility in the folded protein. Biochemistry 26:3754–3758

Anfinsen CB (1973) Principles that govern the folding of protein chains. Science 181:223–230

Anfinsen CB, Scheraga HA (1975) Experimental and theoretical aspects of protein folding. Adv Protein Chem 29:205–300

Arakawa T, Timasheff SN (1982) Preferential interactions of proteins with salts in concentrated solutions. Biochemistry 21:6545–6552

Argos P, Rossmann MG, Grau UM, Zuber H, Frank G, Tratschin JD (1979) Thermal stability of protein structure. Biochemistry 18:5698–5703

Baldwin RL (1986) Temperature dependence of the hydrophobic interaction in protein folding. Proc Natl Acad Sci USA 83:8069–8072

Baldwin RL, Eisemberg D (1987) Protein stability. In: Oxender DL, Fox CF (eds) Protein engineering. Liss, New York, pp 127–148

Barnes LD, Stellwagen E (1973) Enolase from the thermophile *Thermus* X-1. Biochemistry 12:1559–1565

Baross JA, Deming JW (1983) Growth of "black smoker" bacteria at temperatures of at least 250 °C. Nature 303:423–426

Beasty AM, Hurle M, Manz JT, Stackhouse T, Matthews CR (1987) Mutagenesis as a probe of protein folding and stability. In: Oxender DL, Fox CF (eds) Protein engineering. Liss, New York, pp 91–102

Becktel WJ, Schellman JA (1987) Protein stability curves. Biopolymers 26:1859–1877
Bernhardt G, Lüdemann H-D, Jaenicke R (1984) Biomolecules are unstable under "black smoker" conditions. Naturwissenschaften 71:583–585
Biesecker G, Harris JI, Thierry JC, Walker JE, Wonacott AJ (1977) Sequence and structure of D-glyceraldehyde-3-phosphate dehydrogenase from *Bacillus stearothermophilus*. Nature 266:328–333
Brock TD (1978) Thermophilic microorganisms and life at high temperature. Springer, Berlin Heidelberg New York
Brock TD (1985) Life at high temperature. Science 230:132–138
Brock TD (1986) Thermophiles: general, molecular and applied microbiology. Wiley, New York
Bryan PN, Rollence ML, Pantoliano MW, Wood J, Finzel BC, Gilliland GL, Howard AJ, Poulos TL (1986) Proteases with enhanced stability: characterization of a thermostable variant of subtilisin. Proteins 1:326–334
Bull HB, Breese K (1973) Thermal stability of proteins. Arch Biochem Biophys 158:681–686
Chen B, Schellman JA (1989) Low-temperature unfolding of a mutant of phage T4 lysozyme. I. Equilibrium studies. Biochemistry 28:685–691
Chibata I (1978) Immobilized enzymes: research and development. Wiley, New York
Colman PM, Jansonius JN, Matthews BW (1972) The structure of thermolysin: an electron density map at 2.3 Å resolution. J Mol Biol 70:701–724
Cowan D, Daniel RM, Morgan HW (1985) Thermophilic proteases: properties and potential applications. Trends Biotechnol 3:68–72
Creighton TE (1978) Experimental studies of protein folding and unfolding. Prog Biophys Mol Biol 33:231–297
Creighton TE (1985) The problem of how and why proteins adopt folded conformations. J Phys Chem 89:2452–2459
Creighton TE (1988) The protein folding problem. Science 240:343–344
Dahlquist FN, Long JW, Bigbee WL (1976) Role of calcium in the thermal stability of thermolysin. Biochemistry 15:1103–1111
Daniel RM (1986) The stability of proteins from extreme thermophiles. In: Oxender DL (ed) Genex-UCLA Symposium, vol 39. Liss, New York, pp 291–296
Daniel RM, Cowan DA, Morgan HW (1981) The industrial potential of enzymes from extremely thermophilic bacteria. Chem Ind NZ 15:94–97
Daniel RM, Cowan DA, Morgan HW, Curran P (1982) A correlation between protein thermostability and resistance to proteolysis. Biochem J 207:641–644
Deetz JS, Rozzell JD (1988) Enzyme-catalyzed reactions in non-aqueous media. Trends Biotechnol 6:15–19
De Grado WF (1988) Design of peptides and proteins. Adv Protein Chem 39:51–124
Deming JW (1986) The biotechnological future for newly described, extremely thermophilic bacteria. Microbiol Ecol 12:111–119
Dill KA (1985) Theory for the folding and stability of globular proteins. Biochemistry 24:1501–1509
Dill KA (1987) The stabilities of globular proteins. In: Oxender DL, Fox CF (eds) Protein engineering. Liss, New York, pp 187–192
Drucker H, Borchers SL (1971) The role of calcium in thermolysin. Effect on kinetic properties and autodigestion. Arch Biochem Biophys 147:242–248
Estell DA, Grayard TP, Wells JA (1985) Engineering an enzyme by site-directed mutagenesis to be resistant to chemical oxidation. J Biol Chem 260:6518–6521
Fabry S, Hensel R (1987) Purification and characterization of D-glyceraldehyde-3-phosphate dehydrogenase from the thermophilic archaebacterium *Methanothermus fervidus*. Eur J Biochem 165:147–155
Fabry S, Lehmacher A, Bode W, Hensel R (1988) Expression of the glyceraldehyde-3-phosphate dehydrogenase gene form the extremely thermophilic archaebacterium *Methanothermus fervidus* in *E. coli*: enzyme purification, crystallization and preliminary crystal data. FEBS Lett 237:213–217
Fassina G, Vita C, Dalzoppo D, Zamai M, Zambonin M, Fontana A (1986) Autolysis of thermolysin: isolation and characterization of a folded three-fragment complex. Eur J Biochem 156:221–228
Feder J, Garrett LR, Wildi BS (1971) Studies on the role of calcium in thermolysin. Biochemistry 10:4552–4556

Fersht AR, Shi J-P, Wilkinson AJ, Blow DM, Carter P, Waye MMY, Winter GP (1984) Analysis of enzyme structure and activity by protein engineering. Angew Chem Int Ed Engl 23:467–473
Fontana A (1984) Thermophilic enzymes and their potential use in biotechnology. Proc 3rd Eur Congr Biotechnology, vol 1. VCV Weinheim FRG, pp 187–192
Fontana A (1988) Structure and stability of thermophilic enzymes: studies on thermolysin. Biophys Chem 29:181–193
Fontana A (1989) Limited proteolysis of globular proteins occurs at exposed and flexible loops. In: Kotyk A, Skoda J, Pachek C, Kostka V (eds) Highlights in modern biochemistry, vol 2. VSP Zeist, The Netherlands, pp 1711–1726
Fontana A, Boccù E, Veronese FM (1976) Effect of EDTA on the conformational stability of thermolysin. Experientia (Suppl) 26:55–59
Fontana A, Vita C, Boccù E, Veronese FM (1977) A fluorimetric study of the role of calcium in the stability of thermolysin. Biochem J 165:539–545
Fontana A, Fassina G, Vita C, Dalzoppo D, Zamai M, Zambonin M (1986a) Correlation between sites of limited proteolysis and segmental mobility in thermolysin. Biochemistry 25:1847–1851
Fontana A, Fassina G, Vita C, Dalzoppo D, Zamai M, Zambonin M (1986b) Structural and stability features of thermolysin revealed by its autolytic process. In: Bertini I, Luchinat C, Maret W, Zeppezauer M (eds) Zinc enzymes. Birkhäuser Basel, pp 225–237
Fontana A, Vita C, Dalzoppo D, Zambonin M (1989) Limited proteolysis as a tool to detect structure and dynamic features of globular proteins: studies on thermolysin. In: Wittman-Liebold B (ed) Methods in protein sequence analysis. Springer, Berlin Heidelberg New York Tokyo, pp 315–324
Friedman SM (ed) (1978) Biochemistry of thermophily. Academic Press, New York
Frömmel C, Höhne WE (1981) Influence of calcium binding on the thermal stability of "thermitase", a serine protease from *Thermoactinomyces vulgaris*. Biochim Biophys Acta 670:25–31
Frömmel C, Sander C (1989) Thermitase, a thermostable subtilisin: comparison of predicted and experimental structures and the molecular cause of thermostability. Proteins 5:22–37
Fujita SC, Oshima T, Imahori K (1976) Purification and properties of D-glyceraldehyde-3-phosphate dehydrogenase from an extreme thermophile, *Thermus thermophilus* strain HB8. Eur J Biochem 64:57–68
Fukusumi S, Kamizono A, Horinouchi S, Beppu T (1988) Cloning and nucleotide sequence of a heat-stable amylase gene from an anaerobic thermophile (*Dictyoglomus thermophilum*). Eur J Biochem 174:15–21
Garrett RA (1985) The uniqueness of archaebacteria. Nature 318:233–235
Gerlt JA (1987) Relationships between enzymatic catalysis and active site structure revealed by applications of site-directed mutagenesis. Chem Rev 87:1079–1105
Ghélis C, Yon J (1982) Protein folding. Academic Press, New York
Ginsburg A, Carroll WR (1965) Some specific ion effects on the conformation and thermal stability of ribonuclease. Biochemistry 4:2159–2174
Goldberg AL, Dice JF (1974) Intracellular protein degradation in mammalian and bacterial cells. Annu Rev Biochem 43:835–869
Goldenberg DP (1985) Dissecting the roles of individual interactions in protein stability: lessons from a circularized protein. J Cell Biochem 29:321–335
Goldenberg DP (1988) Genetic studies of protein stability and mechanisms of folding. Annu Rev Biophys Biophys Chem 17:481–507
Griko YV, Privalov PL, Sturtevant JM, Venyaminov SY (1988) Cold denaturation of staphylococcal nuclease. Proc Natl Acad Sci USA 85:3343–3347
Grütter MG, Hawkes RM, Matthews BW (1979) Molecular basis of thermostability in the lysozyme from bacteriophage T4. Nature 277:667–669
Guagliardi A, Manco G, Rossi M, Bartolucci S (1989) Stability and activity of a thermostable malic enzyme in denaturants and water-miscible organic solvents. Eur J Biochem 183:25–30
Hartley BS (1986) Commercial prospects for enzyme engineering. Philos Trans R Soc Lond A317:321–331
Hartley BS, Payton MA (1983) Industrial prospects for thermophiles and thermophilic enzymes. Biochem Soc Symp 48:133–146
Hawkes R, Grütter MG, Schellman JA (1989) Thermodynamic stability and point mutations of bacteriophage T4 lysozyme. J Mol Biol 175:195–212

Hecht MH, Sturtevant JM, Sauer RT (1986) Stabilization of repressor against themal denaturation by site-directed Gly Ala changes in α-helix 3. Proteins 1:43–46

Heinrich MR (ed) (1976) Extreme environments: mechanisms and microbial adaptation. Academic Press, New York

Hensel R, Laumann S, Heumann H, Lottspeich F (1987) Characterization of two D-glyceraldehyde-3-phosphate dehycrogenases from the extremely thermophilic archaebacterium *Thermoproteus tenax*. Eur J Biochem 170:325–333

Hirata H, Negoro S, Okada H (1985) High production of thermostable β-galactosidase of *Bacillus stearothermophilus* in *Bacillus subtilis*. Appl Environ Microbiol 49:1547–1549

Holmes MA, Matthews BW (1982) Structure of thermolysin refined at 1.6 Å resolution. J Mol Biol 160:623–639

Huber R, Langworthy TA, König H, Thomm M, Woese CR, Sleytr UB, Stetter KO (1986) *Thermotoga maritima* sp. nor represents a new genus of unique extremely thermophilic eubacteria growing up to 90 °C. Arch Microbiol 144:324–333

Iijima S, Uozumi T, Beppu T (1986) Molecular cloning of *Thermus flavus* malate dehydrogenase gene. Agric Biol Chem 50:589–592

Ikai A (1980) Thermostability and aliphatic index in globular proteins. J Biochem (Tokyo) 88:1895–1898

Imanaka T (1983) Host-vector systems in thermophilic *Bacilli* and their applications. Trends Biotechnol 1:139–144

Imanaka T, Shibagaki M, Takagi M (1986) A new way of enhancing the thermostability of proteases. Nature 324:695–697

Jaenicke R (1981) Enzymes under extremes of physical conditions. Annu Rev Biophys Bioeng 10:1–67

Jaenicke R (1987) Folding and association of proteins. Prog Biophys Mol Biol 49:117–237

Joliff G, Béguin P, Juy M, Ryter A, Poljak R, Aubert J-P (1986) Isolation, crystallization and properties of a new cellulase of *Clostridium thermocellum* overproduced in *Escherichia coli*. Biotechnology 4:896–900

Kandler O (1981) Archaebacteria and phylogeny of organisms. Naturwissenschaften 68:183–192

Kandler O (ed) (1982) Archaebacteria. Fischer, Stuttgart

Kandler O (1984) Archaebacteria: biotechnological implications. Proc 3rd Eur Congr Biotechnology, vol 4. VCH Weinheim FRG, pp 551–560

Karpusas M, Baase WA, Matsumura M, Matthews BW (1989) Hydrophobic packing in T4 lysozyme probed by cavity-filling mutants. Proc Natl Acad Sci USA 86:8237–8241

Klibanov AM, Ahern TJ (1987) Thermal stability of proteins. In: Oxender DL, Fox CR (eds) Protein engineering. Liss, New York, pp 213–218

Kristjansson JK (1989) Thermophilic organisms as sources of thermostable enzymes. Trends Biotechnol 7:349–353

Kubo M, Imanaka T (1988) Cloning and nucleotide sequence of the highly thermostable neutral protease gene from *B. stearothermophilus*. J Gen Microbiol 134:1883–1892

Lamed R, Bayer EA, Saha BC, Zeikus JC (1988) Biotechnological potential of enzymes from unique thermophiles. In: Durand G, Bobichon L, Florent J (eds) 8th Int Biotechnology Symposium, Societé Française de Micróbiologie, pp 371–383

Langridge J (1968) Genetic and enzymatic experiments relating to the tertiary structure of β-galactosidase. J Bacteriol 96:1711–1717

Lapanje S (1978) Physicochemical aspects of protein denaturation. Wiley, New York

Leatherbarrow RJ, Fersht AR (1986) Protein engineering. Protein Eng 1:7–16

Liao H, McKenzie T, Hageman R (1986) Isolation of a thermostable enzyme variant by cloning and selection in a thermophile. Proc Natl Acad Sci USA 83:576–580

Ljungdahl LG, Sherod D (1976a) Proteins from thermophilic microorganisms. In: Heinrich M (ed) Mechanisms of microbial adaptation. Academic Press, New York, pp 147–187

Ljungdahl LG, Sherod D (1976b) Thermophilic microorganisms in extreme environments. Bacteriol Rev 37:320–342

Love DR, Streiff MB (1987) Molecular cloning of a β-glucosidase gene from an extremely thermophilic anaerobe in *E. coli* and *B. subtilis*. Biotechnology 5:384–387

Matsumura M, Aiba S (1985) Screening for thermostable mutant of kanamycin nucleotidyltransferase by the use of a transformation system for a thermophile (*B. stearothermophilus*). J Biol Chem 260:15228–15233

Matsumura M, Yasumura S, Aiba S (1986) Cumulative effect of intragenic amino acid replacements on the thermostability of a protein. Nature 323:356–358

Matsumura M, Becktel WJ, Levitt M, Matthews BW (1989a) Stabilization of phage T4 lysozyme by engineering disulfide bonds. Proc Natl Acad Sci USA 86:6562–6566

Matsumura M, Signor G, Matthews BW (1989b) Substantial increase of protein stability by multiple disulfide bonds. Nature 142:291–293

Matsumura M, Wozniak JA, Dao-pin S, Matthews BW (1989c) Structural studies of mutants of T4 lysozyme that alter hydrophobic stabilization. J Biol Chem 264:16059–16066

Matthews BW (1987) Genetic and structural analysis of the protein stability problem. Biochemistry 26:6885–6888

Matthews BW, Jansonius JN, Colman PM, Schenborn BP, Dupourque D (1972) Three-dimensional structure of thermolysin. Nature 238:37–41

Matthews BW, Weaver LH, Kester WR (1974) The conformation of thermolysin. J Biol Chem 249:8030–8044

Matthews BW, Nicholson H, Becktel WJ (1987) Enhanced protein thermostability from site-directed mutations that decrease the entropy of unfolding. Proc Natl Acad Sci USA 84:6663–6667

Matthews CR, Crisanti MM, Gepner GL, Velicelebi G, Sturtevant J (1980) Effect of single amino acid substitutions on the thermal stability of the α-subunit of tryptophan synthase. Biochemistry 19:1290–1293

Menéndez-Arias L, Argos P (1989) Engineering protein thermal stability: sequence statistics point to residue substitution in α-helices. J Mol Biol 206:397–406

Merkler DJ, Farrington GR, Wedler FC (1981) Protein thermostability. Correlations between calculated macroscopic parameters and growth temperature for closely related thermophilic and mesophilic *Bacilli*. Int J Peptide Protein Res 18:430–442

Miller SL, Bada JL (1988) Submarine hot springs and the origin of life. Nature 334:609–611

Mozhaev VV, Martinek K (1984) Structure-stability relationships in proteins: new approaches to stabilizing enzymes. Enzyme Microb Technol 6:50–59

Mozhaev VV, Berezin IV, Martinek K (1988) Structure-stability relationship in proteins: fundamental tasks and strategy for the development of stabilized enzyme catalysts for biotechnology. CRC Crit Rev Biochem 23:235–281

Mulkerrin MG, Perry LJ, Wetzel R (1986) Stability and solution structure of a disulfide cross-linked T4 lysozyme. In: Oxender DL (ed) Protein structure, folding and design. Liss, New York, pp 297–305

Nagahari K, Koshikawa T, Sakaguchi K (1980) Cloning and expression of the leucine gene from *Thermus thermophilus* in *Escherichia coli*. Gene 10:137–145

Nagata S, Tanizawa K, Esaki N, Sakamoto Y, Ohshima T, Tanaka H, Soda K (1988) Gene cloning and sequence determination of leucine dehydrogenase from *Bacillus stearothermophilus* and structural comparison with other $NAD(P)^+$-dependent dehydrogenases. Biochemistry 27:9056–9062

Nicholson H, Becktel WJ, Matthews BW (1988) Enhanced protein thermostability from designed mutations that interact with α-helix dipoles. Nature 336:651–656

Nojima H, Ikai A, Oshima T, Noda H (1977) Reversible thermal unfolding of thermostable phosphoglycerate kinase. Thermostability associated with mean zero enthalpy change. J Mol Biol 116:429–442

Nosoh Y, Sekiguchi T (1990) Protein engineering for thermostability. Trends Biotechnol 8:16–20

Ogasahara K, Tsunasawa S, Soda Y, Yutani K, Sugino Y (1985) Effect of single amino acid substitutions on the protease susceptibility of tryptophan synthase α-subunit. Eur J Biochem 150:17–21

Ohlendorf DH, Finzel BC, Weber PC, Salemme FR (1987) Some design principles or structurally stable proteins. In: Oxender DL, Fox GF (eds) Protein engineering. Liss, New York, pp 165–173

Oshima T (1979) Molecular basis for unusual thermostabilities of cell constituents from an extreme thermophile, *Thermus thermophilus*. In: Shilo M (ed) Strategies of microbial life in extreme environments. Verlag Chemie, Weinheim, pp 455–469

Oshima T, Soda K (1989) Thermostable amino acid dehydrogenases. Applications and gene cloning. Trends Biotechnol 7:210–214
Oste C (1988) Polymerase chain reaction. Biotechniques 6:162–167
Oxender DL, Fox CF (eds) (1987) Protein engineering. Liss, New York
Pace CN (1975) The stability of globular proteins. CRC Crit Rev Biochem 3:1–43
Pace CN, Laurents DR (1989) A new method for determining the heat capacity change for protein folding. Biochemistry 28:2520–2525
Pantoliano MW, Ladner RC, Bryan PN, Rollence ML, Wood JF, Poulos TL (1987a) Protein engineering of subtilisin BPN': enhanced stabilization through the introduction of two cysteines to form a disulfide bond. Biochemistry 26:2077–2082
Pantoliano MW, Ladner RC, Bryan PN, Rollence ML, Wood JF, Gilliland GL, Stewart DB, Poulos TL (1987b) The engineering of disulfide bonds, electrostatic interactions and hydrophobic contacts for the stabilization of subtilisin BPN'. Protein Eng 1:229–335
Pantoliano MW, Whitlow M, Wood JF, Rollence ML, Finzel BC, Gilliland GL, Poulos T, Bryan PN (1988) The engineering of binding affinity at metal ion binding sites for the stabilization of proteins: subtilisin as a test case. Biochemistry 27:8311–8317
Pantoliano MW, Whitlow M, Wood JF, Dodd SW, Hardman KD, Rollence ML, Bryan PN (1989) Large increase in general stability for subtilisin BPN' through incremental changes in the free energy of unfolding. Biochemistry 28:7205–7213
Pauptit RA, Karlsson R, Picot D, Jenkins JA, Niklas-Reimer AS, Jansonius JN (1988) Crystal structure of neural protease from *Bacillus cereus* refined at 3.0. A resolution and comparison with the homologous, but more thermostable enzyme thermolysin. J Mol Biol 199:525–537
Perry LJ, Wetzel R (1984) Disulfide bond engineered into T4 lysozyme:stabilization of the protein towards thermal inactivation. Science 226:555–557
Perutz MF (1978) Electrostatic effects in proteins. Nature 201:1187–1191
Perutz MF, Raidt H (1975) Stereochemical basis of heat stability in bacterial ferredoxin and in hemoglobin A2. Nature 255:256–259
Pfeil W (1981) The problem of the stability of globular proteins. Mol Cell Biochem 40:3–28
Ponnuswamy PK, Muthusamy R, Manavalan P (1982) Amino acid composition and thermal stability of proteins. Int J Biol Macromol 4:186–191
Privalov PL (1979) Stability of proteins: small globular proteins. Adv Prot Chem 33:167–241
Privalov PL (1982) Stability of proteins: proteins which do not present a single cooperative system. Adv Protein Chem 35:1–104
Privalov PL, Griko YU, Venyaminov SY, Kutyshenko VP (1986) Cold denaturation of myoglobin. J Mol Biol 190:487–498
Rastetter WH (1983) Enzyme engineering: applications and promise. Trends Biotechnol 1:80–84
Rella R, Raia CA, Trincone A, Gambacorta A, De Rosa M, Rossi M (1987) Properties and specificity of an alcohol dehydrogenase from the thermophilic archaebacterium *Sulfolobus solfataricus*. In: Laane C (ed) Biocatalysis in organic media. Elsevier Science, Amsterdam, pp 273–278
Rella R, Pisani FM, Grandi C, Fontana A, Rossi M (1988) Heat- and detergent-induced activation of β-galactosidase from *Sulfolobus solfataricus*. Proc 3rd Meet Proteins. Naples, Italy, 5–7 May Abstr p 48
Remington SJ, Anderson WF, Owen J, Ten Eyck LF, Grainger CT, Matthews BW (1978) Structure of the lysozyme from bacteriophage T4: an electron density map at 2.4 Å resolution. J Mol Biol 118:81–88
Richardson JS (1981) The anatomy and taxonomy of protein structure. Adv Protein Chem 34:167–339
Roche RS, Voordouw G (1978) The structural and functional roles of metal ions in thermolysin. CRC Crit Rev Biochem 5:1–23
Rossi M (1987) Applications and potentials of prokaryotic thermophiles. Biofutur 53:39–42
Rossi M (1988) Enzymes from extreme thermophilic bacteria as models of special catalysts. Proc Eur Conf Biotechnology. Verona, November 7–8, pp 46–50
Saha BC, Zeikus JG (1989) Novel highly thermostable pullulanase from thermophiles. Trends Biotechnol 7:234–239
Saiki RK, Gelfand DH, Stoffel S, Scharf SJ, Higuchi R, Horn GT, Mullis KB, Ehrlich HA (1988) Primer-directed enzymatic amplification of DNA with a thermostable DNA polymerase. Science 239:487–491

Salemme FR (1985) Engineering aspects of protein structure. Ann NY Acad Sci 439:97–106
Schellman C (1986) Proteolysis as a probe for motility in mutant bacteriophage TA lysozymes. Biophys J 49:4932
Schellman JA (1987) The thermodynamic stability of proteins. Annu Rev Biophys Chem 16:115–137
Schellman JA, Hawkes RB (1980) The measurement of protein stability. In: Jaenicke R (ed) Protein folding. Elsevier/North Holland Biomedical Press, Amsterdam, pp 331–343
Schellman JA, Lindorfer M, Hawkes R, Grütter M (1981) Mutations and protein stability. Biopolymers 20:1989–1999
Schwarz WG, Grabnitz F, Staudenbauer WL (1986) Properties of a *Clostridium thermocellum* endoglucanase produced in *Escherichia coli*. Appl Environ Microbiol 51:1293–1289
Sekiguchi T, Suda M, Sekiguchi T, Nosoh Y (1986a) Cloning and DNA homology of 3-isopropylmalate dehydrogenase from thermophilic bacilli. FEMS Microbiol Lett 36:41–45
Sekiguchi T, Ortega-Cesena J, Nosoh Y, Ohashi S, Tsuda K, Kanaya S (1986b) DNA and amino acid sequences of 3-isopropylmalate dehydrogenase of *Bacillus coaqulans*: comparison with enzyme of *Saccharomyces cerevisiae* and *Thermus thermophilus*. Biochim Biophys Acta 867:36–44
Shami EY, Rothstein A, Ramjeesingh M (1989) Stabilization of biologically active proteins. Trends Biotechnol 7:186–190
Shaw WV (1987) Protein engineering: the design, synthesis and characterization of factitious proteins. Biochem J 246:1–17
Singleton R, Amelunxen RE (1973) Proteins from thermophilic microorganisms. Bacteriol Rev 37:320–342
Singleton R, Middaugh CR, McElroy RP (1977) Comparison of proteins from thermophilic and non-thermophilic sources in terms of structural parameters inferred from amino acid composition. Int J Peptide Protein Res 10:39–50
Sonnleitner B (1984) Biotechnology of thermophilic bacteria: growth, products and application. Adv Biochem Eng 3:69–138
Sonnleitner B, Fiechter A (1983) Advantages of using thermophiles in biotechnological processes: expectations and reality. Trends Biotechnol 1:74–80
Stellwagen E, Cronlund MM, Barnes LD (1973) A thermostable enolase from the extreme thermophile *Thermus aquaticus* YT-1. Biochemistry 12:1552–1553
Stetter KO (1982) Ultrathin mycelia-forming organisms from submarine volcanic areas having an optimum growth temperature of 105 °C. Nature 300:258–260
Stetter KO, Lauerer G, Thomm M, Neuner A (1987) Isolation of extremely thermophilic sulfate reducers: evidence for a novel branch of archaebacteria. Science 236:822–824
Strehlow KC, Baldwin RL (1989) Effect of the substitution Ala Gly at each of five residue positions in the C-peptide helix. Biochemistry 28:2130–2133
Sundaram TK, Wright IP, Wilkinson AE (1980) Malate dehydrogenase from thermophilic and mesophilic bacteria: molecular size, subunit structure, amino acid composition, immunochemical homology and catalytic activity. Biochemistry 19:2017–2022
Suzuki Y, Imai T (1982) Abnormally high tolerance against proteolysis of an exo-oligo-1,6-glucosidase from thermophile *Bacillus thermoglucosidicus* RP 1006, compared with its mesophilic counterpart from *Bacillus cereus* ATCC 7064. Biochim Biophys Acta 705:124–126
Tajima M, Uvahe I, Yutani K, Okada H (1976) Role of calcium ions in the thermostability of thermolysin and *Bacillus subtilis* var. *amylosacchariticus* neutral protease. Eur J Biochem 64:243–247
Tanaka T, Kawano N, Oshima T (1981) Cloning of the 3-isopropylmalate dehydrogenase of an extreme thermophile and partial purification of the gene product. J Biochem 89:677–682
Tanford C (1968) Protein denaturation. Adv Protein Chem 23:121–282
Tanford C (1970) Protein denaturation. Adv Protein Chem 24:1–95
Tanford C (1980) The hydrophobic effect. Wiley, New York
Titani K, Hermodson MA, Ericsson LH, Walsh KA, Neurath H (1972) Amino acid sequence of thermolysin. Nature 238:35–37
Torchilin VP, Martinek K (1979) Enzyme stabilization without carriers. Enzyme Microb Technol 1:74–82
Tronrud DE, Monzingo AF, Matthews BW (1986) Crystallographic structural analysis of phosphoramidates as inhibitors and transition-state analogs of thermolysin. Eur J Biochem 157:261–268

Tsuboi M, Ohta S, Nakanishi M (1978) Structural fluctuation of protein and thermophily. In: Friedman M (ed) Biochemistry of thermophily. Academic Press, New York, pp 251–266

Tsukagoshi N, Ihara H, Yamagata H, Ukada S (1984) Cloning and expression of a thermophilic α-amylase gene from *Bacillus stearothermophilus* in *Escherichia coli*. Mol Gen Genet 193:58–63

Ulmer KM (1983) Protein engineering. Science 219:666–671

Veronese FM, Boccù E, Schiavon O, Grandi C, Fontana A (1984) General stability of thermophilic enzymes: studies on 6-phosphogluconate dehydrogenase from *B. stearothermophilus* and yeast. J Appl Biochem 6:39–47

Vihinen M (1987) Relationship of protein flexibility to thermostability. Protein Eng 1:477–480

Vita C, Dalzoppo D, Fontana A (1985) Limited proteolysis of thermolysin by subtilisin: isolation and characterization of a partially active enzyme derivative. Biochemistry 24:1798–1806

Volkin DB, Klibanov AM (1987) Thermal destruction processes in proteins involving cystine residues. J Biol Chem 262:2945–2950

von Hippel PH, Schleich T (1969) The effect of salts on the conformational stability of globular proteins. In: Timasheff SN, Fasman GD (eds) Structure and stability of biological macromolecules. Dekker, New York, pp 417–574

von Hippel PH, Wong K-Y (1965) On the conformational stability of globular proteins: the effects of various electrolytes and nonelectrolytes on the thermal ribonuclease transition. J Biol Chem 240:3909–3923

Voordouw G, Roche RS (1975) The role of bound calcium ions in thermostable proteolytic enzymes. II. Studies on thermolysin, the thermostable protease from *Bacillus thermoproteolyticus*. Biochemistry 14:4667–4673

Voordouw G, Milo C, Roche RS (1976) Role of bound calcium ions in thermostable, proteolytic enzymes. Separation of intrinsic and calcium ion contributions to the kinetic thermal stability. Biochemistry 15:3716–3724

Walker JE (1979) Enzymes from thermophilic bacteria. In: Hofmann E (ed) Proteins: structure, function and industrial applications. Pergamon, Oxford, pp 211–225

Walker JE, Wonacott AJ, Harris JJ (1980) Heat stability of a tetrameric enzyme D-glyceraldehyde-3-phosphate dehydrogenase. Eur J Biochem 108:581–586

Weaver LH, Kester WR, Ten Eyk LF, Matthews BW (1976) The structure and stability of thermolysin. Experientia (Suppl) 26:31–39

Weaver LH, Grey TM, Grütter MG, Anderson DE, Wozniak JA, Dahlquist FW, Matthews BW (1989) High-resolution structure of the temperature-sensitive mutant of phage lysozyme, $Arg^{96} \rightarrow$ His. Biochemistry 28:3793–3797

Wells JA, Powers DB (1986) In vivo formation and stability of engineered disulfide bonds in subtilisin. J Biol Chem 261:6564–6570

Wetzel R (1988) Harnessing disulfide bonds using protein engineering. Trends Biochem Sci 12:478–482

Wetzel R, Perry LJ, Baase WA, Becktel WJ (1988) Disulfide bonds and thermal stability in T4 lysozyme. Proc Natl Acad Sci USA 85:401–405

White RH (1984) Hydrolytic stability of biomolecules at high temperatures and its implication for life at 250 °C. Nature 310:430–432

Wiegel J, Ljungdahl LB (1986) The importance of thermophilic bacteria in biotechnology. CRC Crit Rev Biotechnol 3:39–108

Winter G, Fersht AR (1984) Engineering enzymes. Trends Biotechnol 2:115–119

Woese CR (1987) Bacterial evolution. Microbiol Rev 51:221–271

Woese CR, Magrum LJ, Fox GE (1978) Archaebacteria. J Mol Evol 11:245–252

Wütrich K, Roder H, Wagner G (1980) Internal mobility and unfolding of globular proteins. In: Jaenicke R (ed) Protein folding. Elsevier/North Holland Biomedical Press, Amsterdam, pp 549–564

Yutani K, Ogasahara K, Sugino Y, Matsushiro A (1977) Effect of single amino acid substitution on stability of conformation of a protein. Nature 267:274–275

Yutani K, Ogasahara K, Aoki K, Kakuno T, Sugino Y (1984) Effect of amino acid residues on conformational stability of eight mutant proteins variously substituted at a unique position of the tryptophan synthase α-subunit. J Biol Chem 259:14076–14081

Yutani K, Ogasahara K, Tsujita T, Sugino Y (1987) Dependence of conformational stability on

hydrophobicity of the amino acid residue in a series of variant proteins substituted at a unique position of tryptophan synthase alpha subunit. Proc Natl Acad Sci USA 84:4441–4444
Zaks A, Klibanov AM (1985) Enzyme-catalyzed processes in organic solvents. Proc Natl Acad Sci USA 82:3192–3196
Zale SE, Klibanov AM (1983) On the role of reversible denaturation (unfolding) in the irreversible thermal inactivation of enzymes. Biotechnol Bioeng 25:2221–2230
Zale SE, Klibanov AM (1986) Why does ribonuclease irreversibly inactivate at high temperatures? Biochemistry 25:5432–5444
Zeikus JB (1979) Thermophilic bacteria: ecology, physiology and technology. Enzyme Microb Technol 1:243–252
Zuber H (ed) (1976) Enzymes and proteins from thermophilic microorganisms. Birkhäuser, Basel
Zuber H (1981) Structural and functional aspects of enzyme catalysis. In: Eggerer H, Huber R (eds) Structural and functional aspects of enzyme catalysis. Springer, Berlin Heidelberg New York, pp 114–127

Enzymes from Extreme Thermophilic Bacteria as Special Catalysts: Studies on a β-Galactosidase from *Sulfolobus solfataricus*

M. Rossi[2], R. Rella[1], F. Pisani[1], M.V. Cubellis[2], M. Moracci[1], R. Nucci[1], and C. Vaccaro[1]

1 Introduction

Biotechnological applications of proteins and enzymes are often hampered by their low stability to heat, pH, organic solvents, and proteolysis. With the aid of protein engineering, however, many attempts are being made to improve the operational stability of current commercial enzymes, and, in a more general sense, to establish guidelines for improving the thermostability of proteins and enzymes (Mozhaev et al. 1988).

In this context, attention has been focused on proteins and enzymes found in extreme thermophilic bacteria which grow between 70 and 100 °C and over (Stetter 1986; Brock 1985). Many of these special bacteria belong to the new taxonomic unit of archaebacteria, a group of extremophile microorganisms recently recognized as a new order in addition to eubacteria and eukaryotes (Woese and Wolfe 1985).

Some authors believe that archaebacteria represent the direct descendants of the most ancient living form, since they are found in extreme habitats which are most likely similar to the environmental conditions of primitive earth (Woese 1982). Indeed, they have evolved molecular mechanisms, metabolic pathways, and compounds peculiar to their own kingdom. In addition, organisms able to survive and grow under extreme environmental conditions must possess macromolecules, such as lipids and proteins, which have increased stability, among other peculiar properties.

A complete understanding of the biology of extreme thermophilic microorganisms and their successful strategy of adaptation to such high temperatures first requires the understanding of both their metabolism and the structure and function of their cellular components.

In fact, while thermophilic eubacteria have presumably evolved from mesophilic lines by devising different ways to turn mesophilic macromolecules into more thermophilic ones, the extreme thermophilic bacteria, according to Kandler (1984), "may have invented and further transmitted a unique, most successful principle to construct thermostable proteins allowing them to grow even in boiling water".

[1] Institute of Protein Biochemistry and Enzymology, C.N.R., Via Marconi 10, Naples, Italy.
[2] Dipartimento di Chimica Organica e Biologica, Università di Napoli, Via Mezzocannone 16, Naples, Italy.

Guido di Prisco (Ed)
Life Under Extreme Conditions

It appears that whereas conventional proteins and enzymes are irreversibly inactivated by heat, the enzymes isolated from extreme thermophiles are thermostable and thermophilic, that is, capable of functioning at high temperatures. In addition, they show enhanced stability in the presence of common protein denaturants, such as sodium dodecyl sulfate, urea, guanidine hydrochloride, organic solvent, and proteolytic enzymes. Their stability is further enhanced by immobilization. In contrast, in a number of immobilization procedures developed for the mesophilic enzymes, this thermostability is often an unattained goal (Fontana 1988; Mozhaev et al. 1988).

What is the molecular mechanism responsible for such extraordinary properties? Evidence so far collected indicates that the enhanced stability of a thermophilic enzyme cannot be attributed to one common factor, but is presumably the result of a variety of interactions such as disulfide bonds, hydrophobic interactions, hydrogen and ion bonding, metal binding, etc. Occasionally, stability can also be achieved extrinsically by the addition of suitable ligands such as metals, as in the case of α-amylase and thermolysin (Fontana 1988).

The most direct approach to the understanding of the molecular basis of thermostability is the comparison of the amino acid sequence and the tridimensional structure of proteins with different heat stability patterns. From the few data available on the properties of some of the thermophilic enzymes studied, and from their primary structure, we argue that they are different, with little or no homology, from their corresponding mesophilic enzymes. In fact, they are examples of thermostable, thermophilic, and solvent-resistant proteins whose structure can be used as a natural model for designing proteins obtainable by protein engineering.

The organism used in our studies is *Sulfolobus solfataricus*, an extreme thermophilic archaebacterium growing at 87 °C at pH 3 (De Rosa et al. 1975); enzyme models used have been an alcohol dehydrogenase, a malic enzyme, and a β-galactosidase (Bartolucci et al. 1987; Rella et al. 1987; Pisani et al. 1989). Here, we will summarize the work done on β-galactosidase.

2 Results and Discussion

2.1 Purification of β-Galactosidase

Sulfolobus solfataricus has a constitutive β-galactosidase localized in the cytosol (De Rosa et al. 1980). The level of the enzyme in the cells is approximately the same if the microorganisms are harvested in the middle logarithmic phase or in the early stationary phase of growth, and is not influenced by the presence of lactose or when the lactose is the only carbon source in the standard culture medium (Buonocore et al. 1980).

The interesting properties of the partially purified enzyme, such as thermostability, thermophilicity, and resistance to water-soluble, organic solvents and detergents prompted us to purify the enzyme to homogeneity to study its structure-function relationships (Buonocore et al. 1980). A purification procedure was develeoped and the enzyme was obtained in homogeneous form after about 1000-fold purification (Pisani et al. 1989).

Table 1 summarizes the procedure for the purification of the enzyme; the crucial step was an affinity chromatography analysis, and the homogeneous enzyme had a specific activity of 115–130 unit/mg of protein at 75 °C, using o-nitrophenyl-β-D-galactopyranoside as substrate.

Table 1. Purification of *S. solfataricus* β-galactosidase[a]

	Step	Total Protein (mg)	Total activity (U)	Specific activity (U/mg)	Yield (%)
I.	Crude extract	6500	750	0.11	100
II.	DE52-cellulose chromatography	484	316	0.65	42
III.	pH 4.5 precipitation	268.5	340	1.26	45
IV.	Affinity chromatography	5.5	222	40	29.6
V.	Sepharose-Q chromatography	1.8	210	116	28

[a] The data refer to a typical purification using 50 g frozen bacterial cells as starting material.

2.2 Structure of β-Galactosidase

The molecular mass of the native β-galactosidase estimated either by gel filtration or glycerol density centrifugation was found to be 240 000 ± 8000 daltons. The subunit composition was studied by electrophoresis, incubating the enzyme for 60 min at 90 °C in the presence of 1% SDS to obtain its complete denaturation, and running the sample with appropriate molecular weight markers. A single band with a molecular mass of 60 000 ± 2000 daltons was obtained, indicating that the enzyme was composed of four similar or identical subunits. This result was confirmed by cross-linking the enzyme with bifunctional reagents according to Davies and Stark (1970), and by amino acid analysis. On SDS-PAGE electrophoresis of β-galactosidase treated with dimethyl-suberimidate (DMS), four protein bands were found, corresponding to Mr of 58, 120, 175, 220 kDa as estimated by using DMS-treated aldolase as the molecular weight marker.

From the amino acid composition shown in Table 2, a subunit molecular mass of 60 690 daltons can be estimated on the basis of 12 histidine residues per subunit. The $\frac{(\mathrm{Asx} + \mathrm{Glx})}{(\mathrm{Lys} + \mathrm{Arg})}$ ratio was 2.2 in accordance with the acidic nature of the protein (Ip = 4.5). Only one cysteine group per subunit was found in accordance with the determination of sulfhydryl groups by Ellman's reaction. The average hydrophobic index, calculated according to Bigelow (1967) from the amino acid composition, was 4416 J/residue.

Table 2. Amino acid composition of *S. solfataricus* β-galactosidase[a]

Amino acid	Residues per subunit
Lysine	24.48
Histidine	12
Arginine	29.88
Aspartic acid/aspargine	65.28
Threonine	24[b]
Serine	46.32[b]
Glutamic acid/glutamine	56.28
Proline	23.76
Glycine	58.8
Alanine	35.52
Valine	31.78
Methionine	6.72
Isoleucine	19.44
Leucine	36.96
Tyrosine	31.30
Phenylalanine	21.84
Cysteine	1.3[c]
Tryptophan	17.52[d]

[a] The data refer to the average of four analyses after 24, 48, and 72 h hydrolysis, with the exception of cysteine and tryptophan.
[b] Values extrapolated to zero hydrolysis time.
[c] Determination as cysteic acid.
[d] Determination by the method of Penke et al. (1974).

2.3 Reaction Requirements: Thermophilicity – Thermostability

In contrast to the enzymes from other sources, the activity of this β-galactosidase was not found to be dependent on the presence of monovalent or divalent metal cations. Maximal activity was observed at pH 6.5 in phosphate buffer with β-NpGal as substrate and its specificity (Table 3) was found to be confined to the sugar moiety and to the anomeric character of the linkage, as is the case of many glucosidases (Wallenfels and Weil 1972). Indeed, the inversion of β-glycosidic linkage to the α-configuration rendered the galactoside resistant to the enzymatic hydrolysis, as indicated by the fact that α-NpGal was not a substrate of the enzyme. Moreover, *S. solfataricus* β-galactosidase displayed a broad specificity with a tolerance regarding structural variations of the aglycon moiety (Table 3).

The enzyme was able to act at very high temperatures. The activity showed a continuous increase up to 95 °C, and its value at 30 °C was about 3% of that at 95 °C.

The thermal stability of *S. solfataricus* β-galactosidase was remarkable and comparable with that of other enzymes purified from the same source (Rossi et al. 1986; Bartolucci et al. 1987; Rella et al. 1987), but greater than that reported for

Table 3. Kinetic constants for *S. solfataricus* β-galactosidase[a]

Substrate	K_m (mM)	k_{cat} (min^{-1})
β-ONPG	0.225	5700
α-Lactose	13	114
Methyl β-D-galactopyranoside	23	242
Phenyl β-D-galactopyranoside	2.6	166

[a]The K_m values for *S. solfataricus* were determined by Lineweaver-Burk plots of the data taken from assays carried out at 75 °C for β-ONPG and at 30 °C for the other substrates by using the coupled system β-galactosidase/β-galactose dehydrogenase.

β-galactosidases from mesophilic and thermophilic bacterial sources (Ulrich et al. 1972; Wallenfels and Weil 1972; Cowan et al. 1984). Figure 1 shows the stability of the enzyme at different temperatures. The half-life at 75, 80, and 85 °C was 24, 10, and 3 h, respectively. However, enzyme activity was completely lost after 15 min at 100 °C and 120 min at 95 °C.

The homogeneous enzyme is very stable in the presence of certain water-miscible solvents and detergents, such as sodium dodecylsulfate; studies are in progress on the effect of such agents on both the activity and stability of the protein.

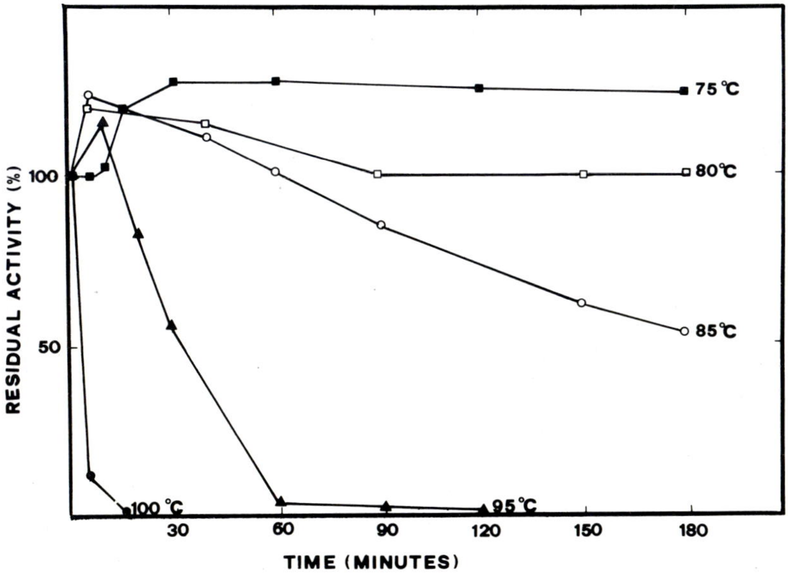

Fig. 1. Thermal stability of *S. solfataricus* β-galactosidase at various temperatures. At the times indicated, aliquots of the preincubated enzyme were withdrawn from the incubation mixtures and assayed at 75 °C under standard conditions

2.4 Cloning, Sequencing, and Expression in *E. coli* of the *β*-Galactosidase Gene

The expression in mesophiles of archaebacterial genes was considered compulsory in order to obtain adequate quantities of enzymes to study them with the appropriate methods (Fabry et al. 1988; Horinouchi et al. 1988). In addition, it was of great interest to fully characterize the expression of genes coding for enzymes from extreme thermophilic bacteria in typical mesophilic hosts, such as *E. coli*. This approach can be useful in understanding the structure and regulation mechanism of these genes and, furthermore, could allow the verification of their expression in heterologous hosts by their own regulation sequences.

β-Galactosidase was chosen as a model because the protein was available in homogeneous form and its main structural and catalytic properties had been studied previously (Pisani et al. 1989). To enable the identification of the gene, polyclonal rabbit antibodies were prepared against the *β*-galactosidase, and the NH_2-terminal sequence of the enzyme was determined (Cubellis et al. 1990).

Two genomic libraries from *S. solfataricus* in λgt 11 and λEMBL 3 vectors were screened, using as probes antibodies and a synthetic oligonucleotide mixture constructed from the NH_2-terminal sequence of the enzyme. A DNA fragment containing the gene coding for *β*-galactosidase and its 5′ and 3′ flanking regions was isolated and completely sequenced.

The nucleotide sequence (Fig. 2) includes a long open reading frame (ORF) from residue 206 to 1696, and an amino acid sequence is encoded starting from ATG at position 229. The NH_2-terminal residues were very similar to those obtained from the purified *β*-galactosidase, suggesting that the first residue of mature protein is also the first translated amino acid. The isolated gene encodes for a protein of 489 amino acids with a predicted molecular mass of 56 650 daltons, which is in agreement with the molecular weight and the amino acid composition determined directly from the purified protein. In addition, the amino acids immediately preceding the termination codon correspond to the carboxy-terminal residues of the purified protein determined by carboxy-peptidase digestion.

As also found for the aspartate aminotransferase gene from *S. solfataricus* (Cubellis et al. 1989), the expression of the *β*-galactosidase gene in *E. coli* was achieved by inserting a DNA fragment of 3 kb gene containing the ORF and its flanking regions in the plasmid pEMBL18 and by transforming a *β*-gal⁻ *E. coli* strain. The expression of the *S. solfataricus β*-gal gene was independent from the orientation of the insert in the plasmid, suggesting that its transcription was not initiated from the vector promoter but from its own regulative sequences.

Fig. 2. Complete nucleotide sequence of *S. solfataricus β*-galactosidase and its flanking regions. The first 33 amino acids whose sequence has been determined on the purified protein are *heavily underlined*. Nucleotide sequences of potential regulatory elements flanking the *β*-galactosidase ORF are *underlined*

```
        10        20        30        40        50        60        70        80        90
AAGGAGAAACTTGGCAGTTTATAACTTGACAGTAGGTTGTGGAGTGATGACTGGATCAATACTAGGAGGAGTAGCATATAATTACGTTAC

       100       110       120       130       140       150       160       170       180
ACAATTTTATAACCCAATATATTCAATAGACCTTATGCTTATCCTATCCTCTATTCTAAGATTCTCGGTATCTCCCCTATTCTTGACCAT

       190       200       210       220       230       240       250       260       270
AAAAGATACTCGCTCAAAGCTTAAATAATATTAATCATAAATAAAGTCATGTACTCATTTCCAAATAGCTTTAGGTTTGGTTGGTCCCAG
                                                  MetTyrSerPheProAsnSerPheArgPheGlyTrpSerGln

       280       290       300       310       320       330       340       350       360
GCCGGATTTCAATCAGAAATGGGAACACCAGGGTCAGAAGATCCAAATACTGACTGGTATAAATGGGTTCATGATCCAGAAAACATGGCA
AlaGlyPheGlnSerGluMetGlyThrProGlySerGluAspProAsnThrAspTrpTyrLysTrpValHisAspProGluAsnMetAla

       370       380       390       400       410       420       430       440       450
GCGGGATTAGTAAGTGGAGATCTACCAGAAAATGGGCCAGGCTACTGGGGAAACTATAAGACATTTCACGATAATGCACAAAAAATGGGA
AlaGlyLeuValSerGlyAspLeuProGluAsnGlyProGlyTyrTrpGlyAsnTyrLysThrPheHisAspAsnAlaGlnLysMetGly

       460       470       480       490       500       510       520       530       540
TTAAAAATAGCTAGACTAAATGTGGAATGGTCTAGGATATTTCCTAATCCATTACCAAGGCCACAAAACTTTGATGAATCAAAACAAGAT
LeuLysIleAlaArgLeuAsnValGluTrpSerArgIlePheProAsnProLeuProArgProGlnAsnPheAspGluSerLysGlnAsp

       550       560       570       580       590       600       610       620       630
GTGACAGAGGTTGAGATAAACGAAAACGAGTTAAAGAGACTTGACGAGTACGCTAATAAAGACGCATTAAACCATTACAGGGAAATATTC
ValThrGluValGluIleAsnGluAsnGluLeuLysArgLeuAspGluTyrAlaAsnLysAspAlaLeuAsnHisTyrArgGluIlePhe

       640       650       660       670       680       690       700       710       720
AAGGATCTTAAAAGTAGAGGACTTTACTTTATACTAAACATGTATCATTGGCCATTACCTCTATGGTTACACGACCCAATAAGAGTAAGA
LysAspLeuLysSerArgGlyLeuTyrPheIleLeuAsnMetTyrHisTrpProLeuProLeuTrpLeuHisAspProIleArgValArg

       730       740       750       760       770       780       790       800       810
AGAGGAGATTTTACTGGACCAAGTGGTTGGCTAAGTACTAGAACAGTTTACGAATTCGCTAGATTCTCAGCTTATATAGCTTGGAAATTC
ArgGlyAspPheThrGlyProSerGlyTrpLeuSerThrArgThrValTyrGluPheAlaArgPheSerAlaTyrIleAlaTrpLysPhe

       820       830       840       850       860       870       880       890       900
GATGATCTAGTGGATGAGTACTCAACAATGAATGAACCTAACGTTGTTGGAGGTTTAGGATACGTTGGTGTTAAGTCCGGTTTTCCCCCA
AspAspLeuValAspGluTyrSerThrMetAsnGluProAsnValValGlyGlyLeuGlyTyrValGlyValLysSerGlyPheProPro

       910       920       930       940       950       960       970       980       990
GGATACCTAAGCTTTGAACTTTCCCGTAGGCATATGTATAACATCATTCAAGCTCACGCAAGAGCGTATGATGGGATAAAGAGTGTTTCT
GlyTyrLeuSerPheGluLeuSerArgArgHisMetTyrAsnIleIleGlnAlaHisAlaArgAlaTyrAspGlyIleLysSerValSer

      1000      1010      1020      1030      1040      1050      1060      1070      1080
AAAAAACCAGTTGGAATTATTTACGCTAATAGCTCATTCCAGCCGTTAACGGATAAAGATATGGAAGCGGTAGAGATGGCTGAAAATGAT
LysLysProValGlyIleIleTyrAlaAsnSerSerPheGlnProLeuThrAspLysAspMetGluAlaValGluMetAlaGluAsnAsp

      1090      1100      1110      1120      1130      1140      1150      1160      1170
AATAGATGGTGGTTCTTTGATGCTATAATAAGAGGTGAGATCACCAGAGGAAACGAGAAGATTGTAAGAGATGACCTAAAGGGTAGATTG
AsnArgTrpTrpPhePheAspAlaIleIleArgGlyGluIleThrArgGlyAsnGluLysIleValArgAspAspLeuLysGlyArgLeu

      1180      1190      1200      1210      1220      1230      1240      1250      1260
GATTGGATTGGAGTTAATTATTACACTAGGACTGTTGTGAAGAGGACTGAAAAGGGATACGTTAGCTTAGGAGGTTACGGTCACGGATGT
AspTrpIleGlyValAsnTyrTyrThrArgThrValValLysArgThrGluLysGlyTyrValSerLeuGlyGlyTyrGlyHisGlyCys

      1270      1280      1290      1300      1310      1320      1330      1340      1350
GAGAGGAATTCTGTAAGTTTAGCGGGATTACCAACCAGCGACTTCGGCTGGGAGTTCTTCCCAGAAGGTTTATATGACGTTTTGACGAAA
GluArgAsnSerValSerLeuAlaGlyLeuProThrSerAspPheGlyTrpGluPhePheProGluGlyLeuTyrAspValLeuThrLys

      1360      1370      1380      1390      1400      1410      1420      1430      1440
TACTGGAATAGATATCATCTCTATATGTACGTTACTGAAAATGGTATTGCGGATGATGCCGATTATCAAAGGCCCTATTATTTAGTATCT
TyrTrpAsnArgTyrHisLeuTyrMetTyrValThrGluAsnGlyIleAlaAspAspAlaAspTyrGlnArgProTyrTyrLeuValSer

      1450      1460      1470      1480      1490      1500      1510      1520      1530
CACGTTTATCAAGTTCATAGAGCAATAAATAGTGGTGCAGATGTTAGAGGGTATTTACATTGGTCTCTAGCTGATAATTACGAATGGGCT
HisValTyrGlnValHisArgAlaIleAsnSerGlyAlaAspValArgGlyTyrLeuHisTrpSerLeuAlaAspAsnTyrGluTrpAla

      1540      1550      1560      1570      1580      1590      1600      1610      1620
TCAGGATTCTCTATGAGGTTTGGTCTGTTAAAGGTCGATTACAACACTAAGAGACTATACTGGAGACCCTCAGCACTAGTATATAGGGAA
SerGlyPheSerMetArgPheGlyLeuLeuLysValAspTyrAsnThrLysArgLeuTyrTrpArgProSerAlaLeuValTyrArgGlu

      1630      1640      1650      1660      1670      1680      1690      1700      1710
ATCGCCACAAATGGCGCAATAACTGATGAAATAGAGCACTTAAATAGCGTACCTCCAGTAAAGCCATTAAGGCACTAAACTTTCTCAAGT
IleAlaThrAsnGlyAlaIleThrAspGluIleGluHisLeuAsnSerValProProValLysProLeuArgHis

      1720      1730      1740      1750      1760      1770      1780      1790      1800
CTCACTATACCAAATGAGTTTTCTTTTAATCTTATTCTAATCTCATTTTCATTAGATTGCAATACTTTCATACCTTCTATATTATTTATT

      1810      1820      1830      1840      1850      1860      1870      1880      1890
TTGTACCTTTTGGGATCTACACTTAATGTTAGCCTAATTGGAAAGTCATTTAGATTTAATACTGTTACCAGTCCATCCCTTTTAATTATT

      1900      1910      1920      1930      1940      1950      1960      1970      1980
AATGAAAATAAGAAGGGATAAGTAGCGATAGCCCTTATTCCGATATGGTCTCCAACAATATCCCTTATTATCTGCCTTGCAACACTAGGG

      1990      2000      2010      2020      2030      2040      2050      2060      2070
TAGAACTCTGAAATCAGATATGGTAGGTAAGTTGTAAGTGATAGGACGTAAACTTTAGAGTTAGAGTAAGTGTTCTGAAAGACTACTGGG

      2080      2090      2100      2110      2120      2130      2140      2150      2160
TGCAATTCGACACCGTTATAGGCGTAAAGGATTGGCGTAGCTCCGTTTAATGAAAATATAGGTCCTACAGGGAAATTGGCTTGCCTCTTG

      2170      2180      2190      2200      2210      2220      2230      2240      2250
TAATATGACCAATAGAACGTTTTCCCATCCCTGGTTAACGCATTGACACTAACACTATCGTAAATCAAGTTACCGACACCAAGAATTTTC

      2260      2270      2280      2290      2300      2310      2320      2330      2340
AGTGCAGTATCCCCCAAGACTTCAATAAGCTTTTTAGCTGCACTTGCTGTAAACATTAAGTTAACTCCCCTATTAAGTAAATCCACAATA

TCTAGA
```

Table 4. Comparison of the native and expressed enzyme

	Native enzyme	Expressed enzyme
K_m	0.23 mM	0.9 mM
Optimal pH	6.5	6.5
Optimal temperature activity	> 100 °C	> 105 °C
Isoelectric point	4.5	4.5
Residual activity after 2 h at 75 °C	100%	100%

A β-galactosidase activity was purified from homogenates of the transformed *E. coli*, which was shown to be as thermophilic and thermostable as the native enzyme (Rossi et al. 1990; Table 4).

3 Conclusion

The amino acid composition of β-galactosidase from *S. solfataricus*, compared with that from *E. coli*, does not indicate remarkable differences in the content of $\frac{(Glx + Asx)}{(Arg + Lys)}$, aromatic residues, and average hydrophobicity, but it shows some characteristics that are indicative for thermophilic proteins, such as a lower cysteine content (1 vs 16 for subunit) and a lower Arg/Lys ratio (1 vs 2.7) with respect to LacZ β-galactosidase. Furthermore, the protein primary structure, inferred from the nucleotide sequence of the gene, exhibits a low degree of similarity when compared to its counterparts from eukaryotes and prokaryotes (Cubellis et al. 1990).

Although these findings have yet to be confirmed with other proteins, the enzymes extracted from extreme thermophilic archaebacteria can be considered to be part of a new class of special catalysts, based on their particular features and structures. These enzymes can be models useful in the elucidation of the molecular bases of enzyme thermophilicity and thermostability.

Acknowledgments. This work was supported by ECC Biotechnology Action Program (contract No. 0052.1) and by Progetto Finalizzato Biotecnologie e Biostrumentazioni, Consiglio Nazionale delle Ricerche (Italy).

References

Bartolucci S, Rella R, Guagliardi AM, Raia CA, Gambacorta A, De Rosa M, Rossi M (1987) Malic enzyme from archaebacterium *Sulfolobus solfataricus*. J Biol Chem 262:7725–7731
Bigelow CC (1967) On the average hydrophobicity of proteins and the relation between it and protein structure. J Theor Biol 16:187–211

Brock TD (1985) Life at high temperatures. Science 230:132–138

Buonocore V, Sgambati O, De Rosa M, Esposito E, Gambacorta A (1980) A constitutive β-Galactosidase from the extreme thermoacidophile archaebacterium *Caldariella acidophila*: properties of the enzyme in the free state and in immobilized whole cells. J Appl Biochem 2:390–397

Cowan DA, Daniel RM, Martin AM, Morgan HV (1984) Some properties of a β-galactosidase from an extremely thermophilic bacterium. Biotechnol Bioeng 26:1141–1145

Cubellis MV, Rozzo C, Marino G, Nitti G, Arnone MI, Sannia G (1989) Cloning and sequencing of gene coding for aspartate aminotransferase from the thermoacidophilic archaebacterium *Sulfolobus solfataricus*. Eur J Biochem 186:375–381

Cubellis MV, Rozzo C, Montecucchi P, Rossi M (1990) Isolation, sequencing and cloning in *Escherichia coli* of a new β-galactosidase archaebacterial gene. Gene (in press)

Davies GE, Stark GR (1970) Use of dimethyl suberimidate, a cross-linking reagent, in studying the subunit structure of oligomeric proteins. Proc Natl Acad Sci USA 66:651–656

De Rosa M, Gambacorta A, Bu' Lock JD (1975) Extremely thermophilic acidophilic bacteria convergent with *Sulfolobus acidocaldarius*. J Gen Microbiol 86:156–164

De Rosa M, Gambacorta A, Nicolaus B, Buonocore V, Poerio E (1980) Immobilized bacterial cells containing a thermostable β-galactosidase. Biotechnol Lett 2:29–34

Fabry S, Lehmacher A, Bode W, Hensel R (1988) Expression of the glyceraldehyde-3-phosphate dehydrogenase gene from the extremely thermophilic archaebacterium *Methanothermus fervidus* in *E. coli*. FEBS Lett 237:213–217

Fontana A (1988) Structure and stability of thermophilic enzymes. Biophys Chem 29:181–183

Horinouchi S, Fukusumi S, Ohshima T, Beppu T (1988) Cloning and expression in *Escherichia coli* of two additional amylase genes of a strictly anaerobic thermophile, *Disctyoglomus thermophilum*, and their nucleotide sequence with extremely low guanine-plus-cytonine contents. Eur J Biochem 176:243–253

Kandler O (1984) Archaebacteria – biotechnological implications. Proc Third European Congr Biotechnology. Symp Futuristic Aspects of Biotechnology, München, vol IV, pp 551–560

Mozhaev VV, Berezin IV, Mantinek K (1988) Structure-stability relationship in proteins: fundamental tasks and strategy for the development of stabilized enzyme catalysts for biotechnology. CRC Crit Rev Biochem 23:235–281

Penke B, Ferenczi R, Kovacs K (1974) A new acid hydrolysis method for determining tryptophan in peptides and proteins. Anal Biochem 60:45–50

Pisani FM, Rella R, Raia CA, Rozzo C, Nucci R, Gambacorta A, De Rosa M, Rossi M (1989) Thermostable β-galactosidase from the archaebacterium *Sulfolobus solfataricus*. Eur J Biochem 187:321–328

Rella R, Raia CA, Pensa M, Pisani FM, Gambacorta A, De Rosa M, Rossi M (1987) A novel archaebacterial NAD^+-dependent alcohol dehydrogenase. Eur J Biochem 167:475–479

Rossi M, Rella R, Pensa M, Bartolucci S, De Rosa M, Gambacorta A, Raia CA, Dell' Aversano Orabona N (1986) Structure and properties of a thermophilic and thermostable DNA polymerase isolated from *Sulfolobus solfataricus*. Syst Appl Microbiol 7:337–341

Rossi M, Cubellis MV, Rozzo C, Moracci M, Rella R (1990) Cloning, sequencing and expression of a new β-galactosidase from the extreme thermophilic *Sulfolobus solfataricus*. In: Jardetsky O, Nicolini C (eds) Protein engineering and structure. Plenum, New York (in press)

Stetter KO (1986) Diversity of extremely thermophilic archaebacteria. In: Brock TD (ed) Thermophiles, vol 7. Wiley, New York, pp 337–341

Ulrich JT, McFeters GA, Temple KC (1972) Induction and characterizaton of β-galactosidase in an extreme thermophile. J Bacteriol 110:691–698

Wallenfels K, Weil R (1972) β-Galactosidase. In: Boyer PD (ed) The enyzmes, 3rd edn, vol 7. Academic Press, New York, pp 617–663

Woese CR (1982) Archaebacteria and cellular origin: an overview. Zentralbl Bakteriol Mikrobiol Hyg I Abt Orig C3:1–17

Woese CR, Wolfe RS (1985) Archaebacteria: the urkingdom. In: Woese CR, Wolfe RS (eds) The bacteria, vol 8. Academic Press, New York, pp 561–564

A Model for the Stabilization of a Halophilic Protein

G. ZACCAI[1] and H. EISENBERG[2]

The process of protein folding from polypeptide to active structure is largely determined by solvent interactions. Proteins with the same function but from different organisms would have evolved different interaction tactics, each appropriate for the particular solvent environment, in order to achieve an identical active site configuration. Enzymes from the extreme halophiles perform the same functions as their counterparts from other organisms but under conditions of extreme salinity; in halobacteria, for example, the cytoplasm is saturated in KCl. In solvents containing molar salt concentrations, "nonhalophilic" proteins are likely to be aggregated, to precipitate, or even to unfold, depending on the type of salt. Halophilic proteins, however, unfold if the solvent salt concentration falls *below* a certain value (2.5 M KCl, for example), still a very high concentration if judged by more usual physiological standards. They are good models for the study of solvent effect contributions to the structure and stability of proteins and of the adaptation mechanisms underlying them.

Structural studies of halophilic proteins have been reviewed by Eisenberg and Wachtel (1987), and Jaenicke (1981) has reviewed studies of enzymes under extreme physical conditions. The influence of *molar* concentrations of salt on soluble proteins are well documented in terms of so-called salting-in and salting-out effects (Von Hippel and Schleich 1969). Salting-out ions stabilize protein structures and favor precipitation. The two effects are related. Consider the salting-out ion as stabilizing the folded protein by making the apolar residues even more insoluble than in water; this will favor a structure in which a maximum of apolar residues are buried, but it will also favor aggregation, and ultimately, precipitation, because of apolar residues on the protein surface which seek to avoid water by intermolecular interactions. Salting-in ions, on the other hand, could be considered as making apolar groups more soluble (perhaps by direct binding to the polypeptide), thus favoring unfolding of the protein. The data on halophilic proteins, however, pose an interesting problem because they are in apparent contradiction with what is expected from the usual salting-out and salting-in behavior of different ions. In this chapter, a recently published model that can account for all experimental observations so far described on the stabilization of halophilic proteins (Zaccai et al. 1989) is reviewed. First, the results of stability measurements as a function of solvent composition on malate

[1](C.N.R.S., U.R.A. 1333), Institut Laue Langevin, 156X, 38042 Grenoble Cedex, France.
[2]Polymer Department, Weizmann Institute of Science, Rehovot 76100, Israel.

Guido di Prisco (Ed)
Life Under Extreme Conditions

dehydrogenase from *Halobacterium marismortui* (hMDH) are described, and it is explained why these results are difficult to interpret. Secondly, the solution structure of hMDH, which is very different in its general features from that of nonhalophilic proteins, is described; and, finally, the stabilization model is presented and discussed.

Because of their sensitivity to salt conditions, halophilic proteins are difficult to fractionate. Malate dehydrogenase was purified in milligram quantities and characterized (Mevarech et al. 1977; Mevarech and Neumann 1977) by using an especially developed technique involving hydrophobic chromatography and affinity gels. The stability curves of this enzyme, as a function of salt type and concentration, are shown in Fig. 1. These measurements were done in the following way: the enzyme was incubated in the salt conditions indicated on the x-axis at a given temperature for a given length of time, after which residual enzymatic activity was assayed under standard conditions. The conjecture that these curves are in fact correlated with the structural stability of the protein in the different solvents is supported by circular dichroism measurements that show almost total loss of secondary structure after incubation in 1 M NaCl (Pundak et al. 1981). Light scattering (Pundak et al. 1981) and neutron scattering (Zaccai et al. 1986b) experiments showed dissociation and concomitant unfolding of the protein (active hMDH is a dimer) in 1 M NaCl, and slow inactivation time constants (several hours) in low salt solvents (Pundak et al. 1981) make it unlikely that the protein is significantly "renatured" in the short time it takes to do the enzymatic assay under standard conditions. The curves in Fig. 1 are similar for NaCl, KCl and NH_4Cl: hMDH is unstable in concentrations of these salts below about 2 M. In $(NH_4)_2SO_4$ and the pH 7 mixture of K_2HPO_4 and KH_2PO_4 (K-phosphate), it is unstable in salt concentrations below about 0.5 M. SO_4^{2-} and phosphate are strongly salting-out ions and Na^+, K^+, Cl^- are fairly neutral in their salting-out behavior (Von Hippel and Schleich 1969), which could explain that the former set of ions is more effective in stabilizing hMDH. The curve for $MgCl_2$, however, is very peculiar. $MgCl_2$ is a salting-in salt, usually favoring unfolding, but in the case of hMDH it stabilizes the protein in a certain concentration range to yield a bell-shaped stabilization curve (Fig. 1c).

The inactivation of hMDH in various solvents is shown in Fig. 2 as a function of time. The protein is significantly more stable in NaCl than in KCl solvents. This is surprising, because the stabilizing or salting-out effects of these two salts are very similar (Von Hippel and Schleich 1969). Stabilization as a function of temperature is shown in Fig. 3. The thermodynamics of protein stabilization depends on a difference in enthalpy between the folded and unfolded state

Fig. 1a-c. Stability of hMDH under different salt concentration conditions. After exposure to the indicated salt concentrations for 9 h at 30 °C, the enzyme activities were measured under standard conditions, i.e., in 4.3 M NaCl, 0.04 M sodium phosphate, pH 7.3, at 25 °C. **a** Salt dependence of the enzymatic activity of hMDH in NH_4Cl, $(NH_4)_2SO_4$, NaCl and KCl (Mevarech and Newmann 1977). **b** Stability of hMDH in KCl and K-phosphate (Zaccai et al. 1989). Incubations were for 24 h at 19 °C. **c** Stability of hMDH in $MgCl_2$ (Zaccai et al. 1989). Incubations were for 24 h at 19 °C

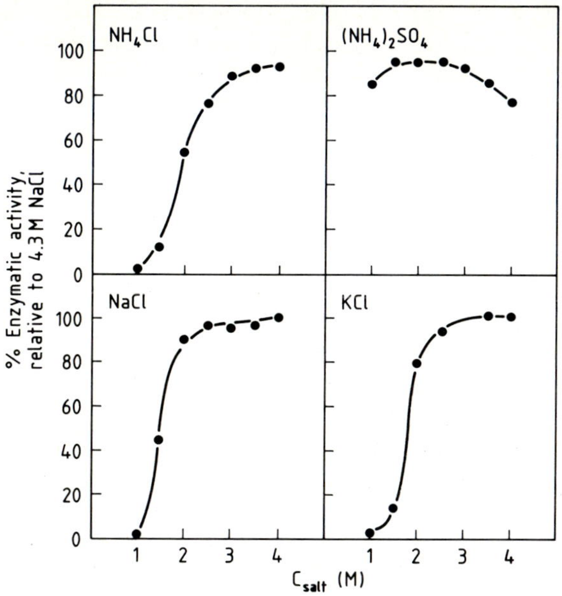
NH4Cl
(NH4)2SO4
NaCl
KCl
% Enzymatic activity, relative to 4.3 M NaCl
Csalt (M)
0
20
40
60
80
100
1
2
3
4

a

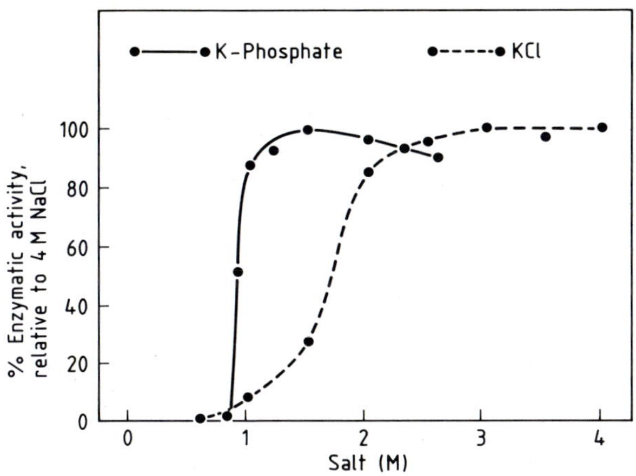
K-Phosphate
KCl
% Enzymatic activity, relative to 4 M NaCl
Salt (M)
0
20
40
60
80
100
0
1
2
3
4

b

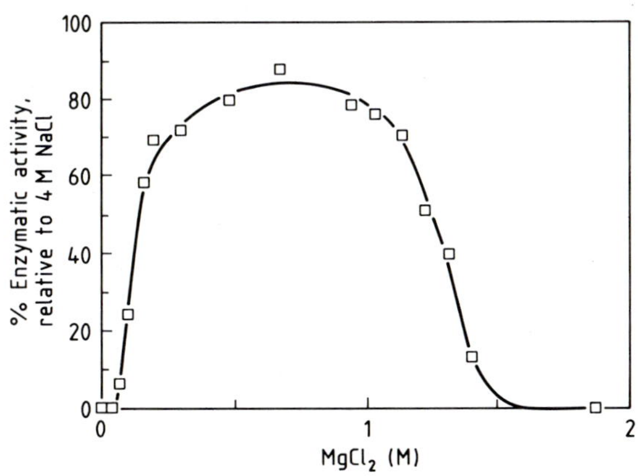
% Enzymatic activity, relative to 4 M NaCl
MgCl2 (M)
0
20
40
60
80
100
0
1
2

c

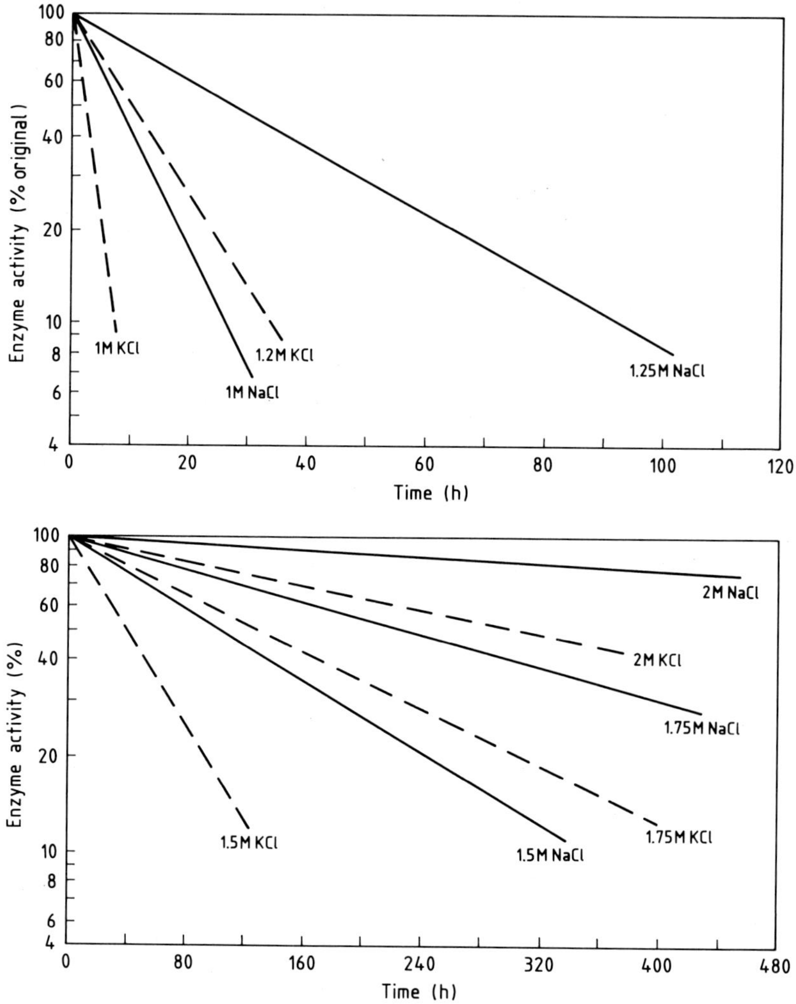

Fig. 2. Inactivation of hMDH in different concentrations of NaCl and KCl at 20 °C. Enzyme activity was measured under standard conditions (as in Fig. 1) after incubation in the salt medium for the time given on the x-axis. (After Pundak et al. 1981)

(arising, for example, from a different number of hydrogen bonds) and a difference in entropy. Privalov (1979) has made an extensive calorimetric study of soluble globular proteins. The difference in enthalpy and the difference in entropy are both temperature-dependent. Close to room temperature, the difference in enthalpy is very small and the entropy term is dominated by the water contribution. An important component of the hydrophobic effect is supposed to be the decrease in entropy "inflicted" on water molecules when nonpolar protein residues are exposed to solvent. Privalov (1979) found a maximum in the

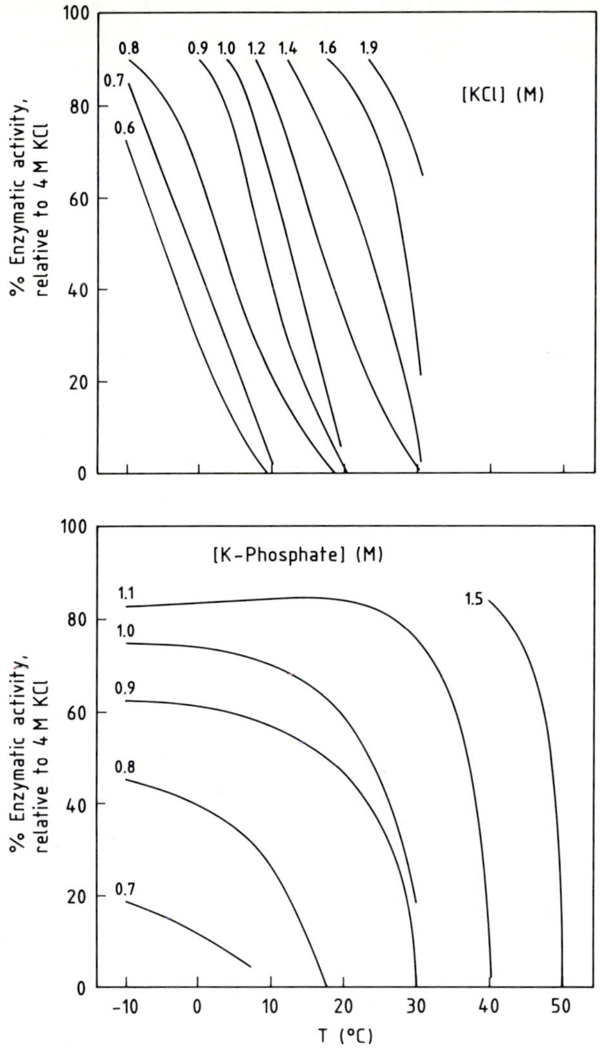

Fig. 3. Enzyme activity of hMDH in standard buffer (Fig. 1) after incubation for 24 h at the salt concentration given, as a function of temperature (Zaccai et al. 1989)

stabilization-free energy close to room temperature, arising from a compensation between the entropy term, which favors the folded structure with increasing temperature, and the enthalpy term, which is destabilizing with increasing temperature. The curves for KCl (Fig. 3) do not have a plateau, suggesting a domination of the enthalpy term at these temperatures. The plateaus in K-phosphate, however, suggest that in this salt, hMDH behaves like a non-halophilic globular protein.

Studies of the solution structure of hMDH yielded results that are quite different from those found for nonhalophilic soluble proteins. The solution structure of a protein is made up of the folded polypeptide chain and a number of solvent molecules with which it is "associated". Eisenberg (1981) has described how the solvent interactions of a macromolecule in solution can be measured, and developed a unified approach for the interpretation of hydrodynamic and scattering data. The key equation is that for the mass (or scattering) density increment of a solution due to the presence of macromolecule (component 2) at constant chemical potential of solvent components (1:water; 3:salt).

$$\partial\rho/\partial c_2 = 1 + \xi_1 - \rho^o (\bar{v}_2 + \xi_1 \bar{v}_1) \ldots \tag{1}$$

where ρ (g/cm^3) is the density of the solution, c_2 (g/cm^3) is the concentration of macromolecule in the solution, ρ^o is the density of the solvent, $\bar{v}_2$ and $\bar{v}_1$ are the partial specific volumes of macromolecule and water, respectively, and ξ_1 is an interaction parameter (gram of water per gram of protein). Because of solvent interactions, the chemical potential of solvent components will change upon addition of the macromolecule; if the solution were dialyzed against a large volume of solvent (of the original composition), ξ_1 would be the mass of water (per gram of macromolecule) that enters the dialysis bag in order to reestablish the initial values for the chemical potentials of water and salt in the solvent. For example, if 1 g of macromolecule bound B_1 g of water and B_3 g of salt, $\xi_1 = B_1 - B_3/w_3$, where the solvent composition is w_3 g of salt per gram of water. Equation (1) can equally well be written in terms of ξ_3 and $\bar{v}_3$, corresponding values for the salt component, with ξ_1 and ξ_3 related by $\xi_1 = -\xi_3/w_3$.

In a radiation scattering experiment, the forward scattered intensity from a solution is proportional to the square of its *scattering density increment*, which is written in a very similar way to Eq. (1). For example, for neutron scattering:

$$\partial\rho_N/\partial c_2 = b_2 + b_1\xi_1 - \rho_N^o (\bar{v}_2 + \xi_1 \bar{v}_1) \ldots \tag{2}$$

where the subscript N denotes neutron scattering length densities, b_2 and b_1 are values of neutron scattering length per gram of macromolecule and per gram of water, respectively. There is a similar equation for X-ray scattering with electron densities in the place of neutron scattering length densities. In the example of the macromolecule with bound water and salt molecules, it can be shown that:

$$\begin{aligned}\partial\rho/\partial n_2 = {} & M_2 + N_1M_1 + N_3M_3 \\ & - \rho^o (\bar{v}_2M_2 + \bar{v}_1M_1N_1 + \bar{v}_3M_3N_3) \ldots \end{aligned} \tag{3}$$

$$\begin{aligned}\partial\rho_N/\partial n_2 = {} & M_2b_2 + N_1M_1b_1 + N_3M_3b_3 \\ & - \rho_N^o (\bar{v}_2M_2 + \bar{v}_1M_1N_1 + \bar{v}_3M_3N_3) \ldots \end{aligned} \tag{4}$$

where macromolecular concentration is now expressed in terms of n_2 mol/cm^3 and bound water and salt in terms of N_1 and N_3 mol/mol of macromolecule, respectively. M_i is the molecular mass (g/mol) of component i. There are trivial relations between these values and the corresponding ones of Eqs. (1) and (2): $n_2 = c_2/M_2$; $B_1 = N_1M_1/M_2$; $B_3 = N_3M_3/M_2$.

Pundak and Eisenberg (1981) first measured values for the solvent interactions of hMDH and found that, in contrast to nonhalophilic globular proteins under similar conditions ($B_1 \sim 0.2$ g/g, $B_3 \sim 0$ g/g), the halophilic protein bound extraordinary amounts of water and salt ($B_1 \sim 0.8$ g/g, $B_3 \sim 0.3$ g/g). Zaccai et al. (1986a), Calmettes et al. (1987), and Zaccai et al. (1989) have exploited the complimentarity of Eqs. (3) and (4) in order to derive B_1 and B_3 (or N_1, N_3) and $\bar{v}_2$ from mass density increment data (obtained from sedimentation and diffusion experiments) and neutron scattering density increment data (from small angle neutron scattering experiments) as a function of the corresponding solvent density. Some of these data are shown in Fig. 4. When B_1 and B_3 are constant, Eqs. (3) and (4) show that the density increment versus solvent density relation is a straight line of gradient $M_2V_T = (\bar{v}_2M_2 + \bar{v}_1M_1N_1 + \bar{v}_3M_3N_3$; the same for mass density and neutron scattering density measurements) and intercept $(M_2 + N_1M_1 + N_3M_3)$ and $(b_2M_2 + b_1M_1N_1 + b_3M_3N_3)$ for mass density and neutron scattering density, respectively. The data fell on straight lines (Fig. 4) and it was possible, therefore, to calculate N_1, N_3, and V_T, thus defining the solution structure of the hMDH particle for each salt condition. The data fell on different lines for the different salts (Fig. 4). For NaCl and KCl, they are very close, showing the solution structure of hMDH to be similar in these two salts. The line for $MgCl_2$, on the other hand, is different in both gradient and intercept, and the point measured for K-phosphate does not fall on any of the lines for the other salts. hMDH, therefore, forms a solution particle of different composition depending on its solvent environment. The parameters calculated from the data are given in Table 1. It can be seen that only in K-phosphate does hMDH have a solution structure similar to a nonhalophilic globular protein. In $MgCl_2$, it binds as many water molecules as in NaCl or KCl, but significantly fewer salt molecules.

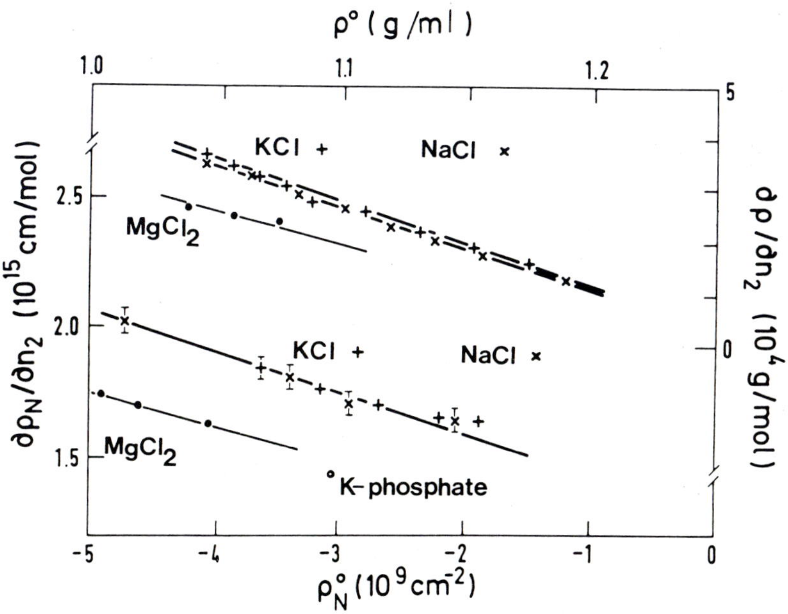

Fig. 4. Mass (*top* and *right-hand y-axis*) and neutron scattering length (*bottom* and *left-hand y-axis*) density increments for hMDH as functions of the respective solvent densities (Zaccai et al. 1989)

Table 1. Composition parameters of the hMDH solution particles, calculated from neutron scattering and mass density increments (Zaccai et al. 1989)

hMDH in	KCl (1–4 M)	$MgCl_2$ (0.5–1.0 M)	K-phosphate (1.5 M)
Protein (M_2 g/mol)	87 000	87 000	87 000
Particle specific volume (V_{tot} cm^3/g)	1.79	1.71	—
Particle volume ($Å^3$) ($M_2 V_{tot}/N_A$)	259 000	247 000	166 000[a]
Protein specific volume ($\bar{v}_2$ cm^3/g)	0.76 ±0.01	0.74 ±0.01	0.75[b]
"Bound" water N_1 (mol water/mol protein)	4080 ±400	4000 ±400	2000 ±1000
"Bound" salt N_3 (mol salt/mol protein)	520 ±60	120 ±30	negligible
Molality of "bound" salt $(N_3 \times 1000)/(N_1 \times M_1)$	7.2 M	1.6 M	—

[a] The sum of the protein and hydration volumes.
[b] Since mass density increments could not be determined for K-phosphate solutions, a value of $v_2 = 0.75$ was assumed in order to interpret the neutron scattering density increment.

An interesting and important observation is that the exceptional solvent binding of hMDH is associated with its quaternary structure (Pundak et al. 1981; Zaccai et al. 1986b). The experiments on the dissociated, unfolded protein in 1 M NaCl are consistent with ~ 0.2 g of water and *no* salt associated with 1 g of protein.

The X-ray and neutron scattering curves of hMDH (Reich et al. 1982; Zaccai et al. 1986a; Calmettes et al. 1987) showed the solution particle to be globular with a relatively large surface to volume ratio (Fig. 5). The schematic structure model presented by Zaccai et al. (1986a) results from the scattering curves to large angles and the radius of gyration values as a function of the contrast between the particle and the solvent. Such "contrast variation" experiments have allowed an estimate of the radius of gyration of the protein moiety of the particle and of the radius of gyration of the bound water and salt moiety. The model has a protein core similar to the shape of pig heart malate dehydrogenase (MDH) with loops extending outwards on which the solvent interactions take place. These loops provide the larger surface area of hMDH when compared to MDH. Before the stabilization model is described, it is useful to recall the observations which it must explain:

1. hMDH is unstable in all salt solvents below a certain salt concentration (its value depending on the salt).
2. Salting-in $MgCl_2$ stabilizes hMDH when the salt concentration is between ~ 0.3 and ~ 1.3 M.
3. The halophilic protein is significantly more stable in NaCl than in KCl.
4. The curves representing stability as a function of temperature are different in KCl and K-phosphate.

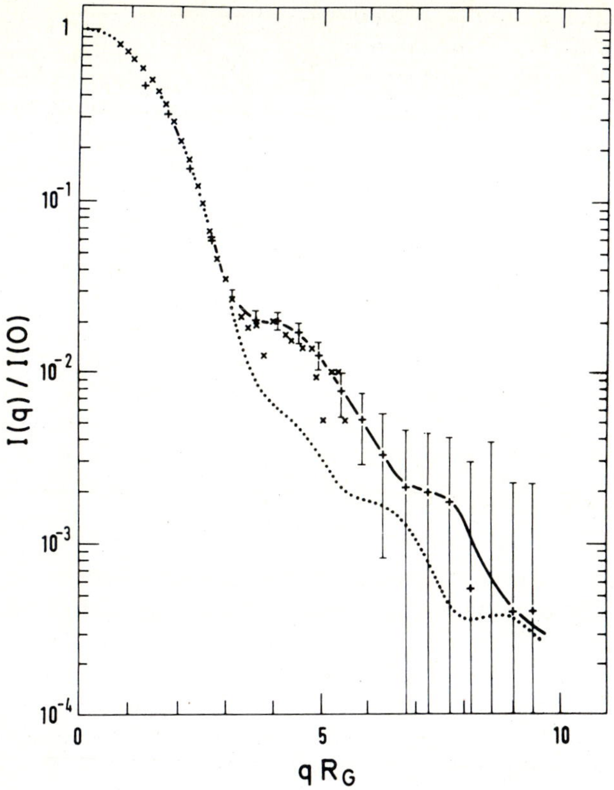

Fig. 5. Normalized neutron scattering intensity of hMDH in 3.5 M KCl in D_2O (+) and in 3.8 M KCl in H_2O (x) as a function of qR_g, where $q = (4\pi \sin\theta)/\lambda \cdot 20$ is the scattering angle and λ the neutron wavelength , and R_g is the scattering radius of gyration of the particle in solution. (......) Scattering curve of an oblate spheroid of axial ratio 1:1:0.6; this is very similar to the scattering curve of nonhalophilic MDH. The higher shoulder in the hMDH curve at $qR_g \sim 4$ is indicative of a higher surface to volume ratio (Calmettes et al. 1987)

5. A relatively large number of water and salt molecules are associated with the tertiary or quaternary structure of hMDH in NaCl and KCl. The solvation is different (but still large) in $MgCl_2$; in K-phosphate, it is similar to that of nonhalophilic globular proteins.

The stabilization model of Zaccai et al. (1989) is based on the fact that the hMDH solution particles differ in the different solvents in which it is active, and on the reasonable assumption that the bound water and salt molecules are not associated separately with the polypeptide but as *hydrated salt ions.*

At low salt, the protein is unfolded because the hydrophobic interaction in these conditions is not sufficient to stabilize the folded structure. If the hydrophobic interaction were strong at low salt (as in a nonhalophilic protein, for example), it is likely that the halophilic protein would precipitate at its physiological concentrations of KCl even though this salt is only mildly salting-out. In molar concentrations of strongly salting-out K-phosphate, however, the hydrophobic interaction is sufficient to stabilize the folded protein; at very high concentrations of this salt, it aggregates and precipitates (Harel et al. 1988). The hydration of hMDH in K-phosphate is similar to that of a nonhalophilic protein. Zaccai et al. (1989) suggested, therefore, that hMDH at low salt and in K-phosphate could be understood in terms of the hydrophobic interaction dominating its stabilization, but with a shift toward higher concentrations of salting-out ions when compared to nonhalophilic globular proteins. The temperature dependence of stabilization in K-phosphate (Fig. 3) is in agreement with this description, the curves with the plateaus near room temperature showing a dominance of the entropy term of the hydrophobic interaction.

The picture for hMDH in solvents with NaCl, KCl or $MgCl_2$ is qualitatively different from that in strongly salting-out conditions (Fig. 6). The model shows that the protein in NaCl, KCl, and $MgCl_2$ solvents is stabilized by the formation of the solution complexes of Table 1. The hMDH dimer in NaCl or KCl interacts with about 500 ions of Na^+ and about 500 of Cl^- hydrated by about 4000 water molecules. In $MgCl_2$, the complex contains approximately the same number of water molecules but much fewer salt ions. This is in agreement with the proposal that solvation is via hydrated ions, because Mg^{++} can coordinate more water molecules than either Na^+ or K^+ (Enderby et al. 1987). Halophilic proteins are particularly rich in acidic residues (Eisenberg and Wachtel 1987), and carboxyl groups are the most solvated of the polar groups in amino acids. Even so, the number of carboxyl groups in the protein is not sufficient to account for the solvation values observed, and it must be assumed that the interaction of the protein with hydrated salt ions is cooperative and related to its tertiary or quaternary structure. This is exactly what is observed; when the protein is dissociated and unfolded, it loses its exceptional solvation properties. The model can account, at least qualitatively, for all the experimental observations on hMDH. Point 5, above, has already been dealt with. The temperature dependence curves (point 4 and Fig. 3) suggest that enthalpy changes dominate stabilization in KCl. The breaking of the chemically coordinated protein-salt-water complex would, in fact, lead to a positive enthalpy change. Point 3 (NaCl stabilizes "better" than KCl) is simply a consequence of the hydration interactions of Na^+ being stronger than those of K^+ (Enderby et al. 1987). Points 1 and 2 are understood in terms of competition between the protein and solvent salt for water molecules. Zaccai et al. (1989) argue that despite its composition, rich in acidic groups, the protein alone would be unable to compete with essentially saturated salt solvents, so it has evolved a tertiary and quaternary structure that can compete effectively by coordinating hydrated salt *at a higher local concentration than in the solvent*. By coordinating hydrated KCl with a local concentration of 7.2 molal (Table 1), the protein is quite safe in its physiological environment (~ 4 M KCl). At low solvent salt concentrations, however, the chemical potential of the salt in the solvent is too low for the complex to form. The implication of this type of argument is that the complex would not form if the solvent salt concentration were *higher* than the

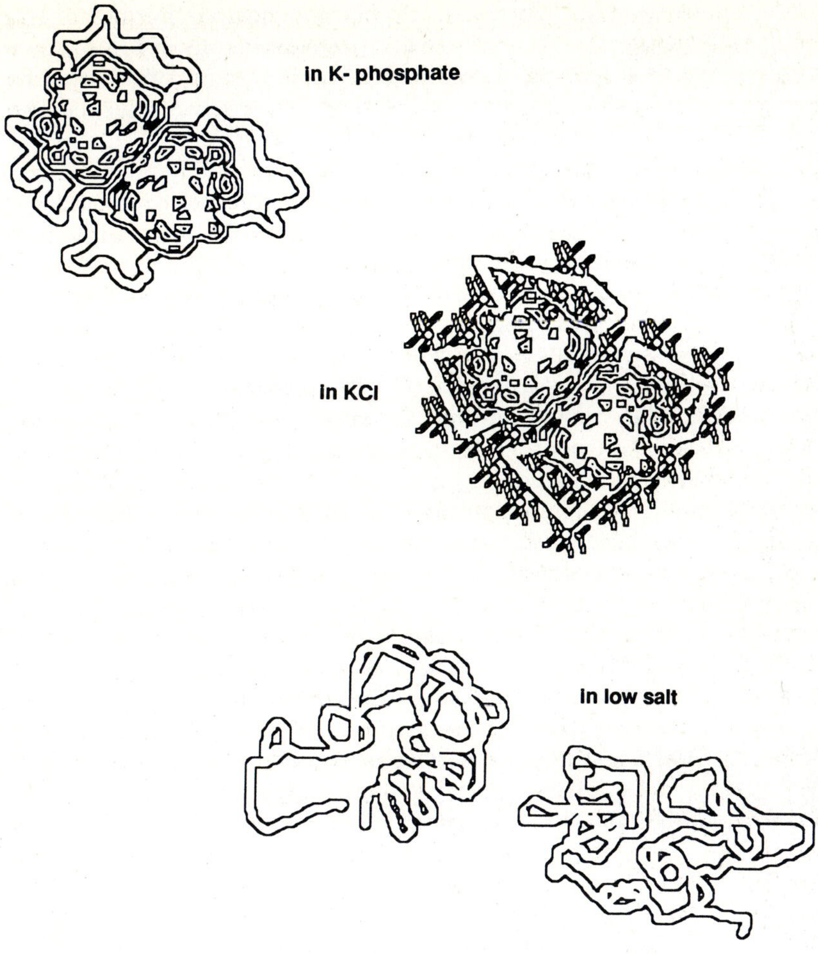

Fig. 6. Schematic representation of hMDH solution structures. The active structures have two parts: a catalytically active core, conceivably similar to that in nonhalophilic MDH, and protruding loops, required for stabilization in KCl, NaCl, or $MgCl_2$ solvents. In *K-phosphate*, the protein dimer is stabilized by the hydrophobicity of the core and the protruding loops are disordered. In *KCl* (or NaCl), the protein is stabilized by the interaction of the loops in a specific protein-water-salt hydration network. In $MgCl_2$, a similar structure exists with the same amount of water molecules coordinated by fewer salt ions. In *low salt*, the protein is unfolded and its hydration is like that of nonhalophilic proteins (Zaccai et al. 1989)

concentration within the complex. This "symmetric" instability at high salt is what is observed in $MgCl_2$, which is soluble beyond 1.6 M, the concentration within the complex (Table 1). In this salt, hMDH is unstable both below and above certain concentrations of the salt (Fig. 1). The higher threshold is similar to the "internal" salt concentration in the complex. It is the behavior of hMDH in nonphysiological $MgCl_2$, therefore, which best illustrates the hypothesis for the stabilization of the halophilic protein in its physiological environment. The models for hMDH in different solvents are drawn schematically in Fig. 6. Since the protein is active in K-phosphate and in KCl (Zaccai et al. 1989) it can be assumed that the active site of hMDH is identical in these two solvents. The protein, therefore, could be considered to be made up of two parts: (1) a constant active site and (2) a part of the structure which can adapt to different environments and which is necessary for the stabilization of the constant part. An interesting view is that hMDH itself behaves like a family of proteins from different organisms, but with the same structure that may have different structural features related to their environments but a constant active site. The difference is that, with hMDH, it is the same primary structure that yields different tertiary structures.

At the end of their report, Zaccai et al. (1989) make suggestions for testing their model further. The primary and tertiary structures of hMDH should be solved. These studies are still in progress. Is the model applicable to all halophilic proteins? There are not sufficient data, yet, to say for sure. It appears, however, that some of the smaller proteins examined from halophilic organisms do not have the exceptional salt and water binding properties of hMDH (H. Eisenberg, unpublished data). The elongation factor Tu from *H. marismortui* (hEF-Tu), of molar mass ~ 43K g/mol, has been purified (Guinet et al. 1988) and preliminary studies have shown it to have stabilization and solution structure behaviour similar to hMDH. Its gene structure has been obtained recently (Baldacci et al. 1990). The analysis of the protein sequence and the comparison with the sequence and structure of *E. coli* EF-Tu have shown that acidic residues that are far apart in the sequence are likely to come together in patches on the surface of the folded halophilic protein. If the model were correct, these residues could cooperatively coordinate the hydrated ion-solvation network.

References

Baldacci G, Guinet F, Tillit J, Zaccai G, de Recondo AM (1990) Functional implications related to the gene structure of the elongation factor EF-Tu from *Halobacterium marismortui*. Nucleic Acids Res 18:507–511

Calmettes P, Eisenberg H, Zaccai G (1987) Structure of halophilic malate dehydrogenase in multimolar KCl solutions from neutron scattering and ultracentrifugation. Biophys Chem 26:279–290

Eisenberg H (1981) Forward scattering of light X-rays and neutrons. Q Rev Biophys 14:141–172

Eisenberg H, Wachtel EJ (1987) Structural studies of halophilic proteins, ribosomes, and organelles of bacteria adapted to extreme salt concentrations. Ann Rev Biophys Biophys Chem 16:69–92

Enderby JE, Cummings S, Herdman GJ, Neilson GW, Salmon PS, Skipper N (1987) Diffraction and the study of aqua ions. J Phys Chem 91:5851–5858

Guinet F, Rainer F, Leberman R (1988) Polypeptide elongation factor Tu from *Halobacterium marismortui*. Eur J Biochem 172:687–694

Harel M, Shoham M, Frolow F, Eisenberg H, Mevarech M, Yonath A, Sussman JL (1988) Crystallization of halophilic malate dehydrogenase from *Halobacterium marismortui*. J Mol Biol 200:609–610

Hippel PH von, Schleich T (1969) The effects of neutral salts on the structure and conformational stability of macromolecules in solution. In: Timasheff SN, Fasman GD (eds) Structure and stability of biological macromolecules. Marcel Dekker, New York, pp 417–574
Jaenicke R (1981) Enzymes under extremes of physical conditions. Annu Rev Biophys Bioeng 10:1–67
Mevarech M, Neumann E (1977) Malate dehydrogenase from extremely halophilic bacteria of the Dead Sea. 2. Effect of salt on the catalytic activity and structure. Biochemistry 16:3786–3791
Mevarech M, Eisenberg H, Neumann E (1977) Malate dehydrogenase isolated from extremely halophilic bacteria of the Dead Sea. 1. Purification and molecular characterization. Biochemistry 16:3781–3785
Privalov PL (1979) Stability of proteins. Small globular proteins. Adv Prot Chem 33:167–241
Pundak S, Eisenberg H (1981) Structure and activity of malate dehydrogenase from the extreme halophilic bacteria of the Dead Sea. 1. Conformation and interaction with water and salt between 5M and 1M NaCl Concentration. Eur J Biochem 118:463–470
Pundak S, Aloni H, Eisenberg H (1981) Structure and activity of malate dehydrogenase from the extreme halophilic bacteria of the Dead Sea. 2. Inactivation, dissociation and unfolding at NaCl concentrations below 2M salt. Salt concentration and temperature dependence of enzyme stability. Eur J Biochem 118:471–477
Reich MH, Kam Z, Eisenberg H (1982) Small-angle X-ray scattering study of halophilic malate dehydrogenase. Biochemistry 21:5189–5195
Zaccai G, Wachtel E, Eisenberg H (1986a) Solution structure of halophilic malate dehydrogenase from small-angle neutron and X-ray scattering and ultracentrifugation. J Mol Biol 190:97–106
Zaccai G, Bunick GJ, Eisenberg H (1986b) Denaturation of a halophilic enzyme monitored by small-angle neutron scattering. J Mol Biol 192:155–157
Zaccai G, Cendrin F, Haik Y, Borochov N, Eisenberg H (1989) Stabilization of halophilic malate dehydrogenase. J Mol Biol 208:491–500

Subject Index